Magic Bullets, Lost Horizons

Magic Bullets, Lost Horizons

THE RISE AND FALL OF ANTIBIOTICS

Sebastian G.B. Amyes

Department of Medical Microbiology
University of Edinburgh
Edinburgh, UK

CRC Press
Taylor & Francis Group
Boca Raton   London   New York

CRC Press is an imprint of the
Taylor & Francis Group, an **informa** business

First published 2001
by Taylor & Francis
11 New Fetter Lane, London EC4P 4EE

Simultaneously published in the USA and Canada
by Taylor & Francis Inc,
29 West 35th Street, New York, NY 10001

Taylor & Francis is an imprint of the Taylor & Francis Group

Typeset in by Expo, Malaysia
Printed and bound in Great Britain by TJ International Ltd,
Padstow, Cornwall

Every effort has been made to ensure that the advice and information
in this book is true and accurate at the time of going to press. However,
neither the publisher nor the authors can accept any legal responsibility
or liability for any errors or omissions that may be made. In the case of
drug administration, any medical procedure or the use of technical
equipment mentioned within this book, you are strongly advised to
consult the manufacturer's guidelines.

British Library Cataloguing in Publication Data
A catalogue record for this book is available from the British Library

Library of Congress Cataloging in Publication Data

ISBN
0-415-27203-3 (hbk)
0-415-27204-1 (pbk)

To Rupert and Alix

"The application of antimicrobial chemotherapy would be the principal cause of a 10 year extension of life expectancy from birth. The magnitude of that change may be taken from the fact that today the elimination of all deaths from cancer would result in only a 2 year extension of life expectancy from birth."

William McDermott with David E. Rogers, "Social ramifications of control of microbial disease". *The Johns Hopkins Medical Journal* **151**: 302–313, 1982.

Contents

Acknowledgements

I should like to thank all my colleagues who have allowed me to pick their brains and engage in long hours of discussion. In particular, I thank Hilary-Kay Young for both philosophical discussions about the use of antibiotics and for some hard facts about the molecular biology of resistance mechanisms. I should also like to thank my research team; firstly for their excellent research results, some of which are included here, and secondly for putting up with absences while this book was being prepared.

I have tried to write a general book on antibiotics that covers both the ethos of their use and the problems that we face. I have used innumerable sources of information, including our university library, the Internet and Public Databases. They are too numerous to list; however I would particularly like to acknowledge the excellent book by Frank Ryan, "Tuberculosis – the Greatest Story Never Told" for the wealth of information that it contained on the historical development of antibiotics, which proved so useful in the preparation of the first two chapters.

Magic bullets

As we enter the third millennium, we take our capability to cure infectious diseases for granted. We anticipate that when we are afflicted by a sore throat, a wheezing cough or an infected cut, there will be a magic pill that will remove the poisonous substance from our bodies. Certainly we have developed an arsenal of antibiotics which can kill or inhibit many of the bacteria that can cause infection and, for the latter part of the twentieth century, the fear that these infections cause has disappeared. We have been much less successful in developing drugs that work against fungal infections and have only limited success against infections caused by viruses. We are still terrified of that one infectious disease that is poised to eradicate the human race, the one against which we have no defence. An epidemic of plague killed an estimated 100 million people in the Middle East, Europe and Asia during the sixth century. Plague returned during the fourteenth century, when it was known as the Black Death, and it is estimated to have killed up to half the population of Europe, or about 75 million people. In the 1980s, HIV appeared poised to cause an AIDS epidemic and in the 1990s, a prion, the causative agent of BSE, seems ready to cause havoc through the population of the United Kingdom. The AIDS epidemic did not materialise, at least not in the developed world, and we simply do not know enough about prion diseases to know if they are equipped to cause a human epidemic of plague proportions. Ignorance has been our greatest handicap and we are still woefully ignorant of many infectious diseases.

In the distant past, the major causes of premature death were precipitated by some infection or another. These were not necessarily epidemics but the general background of infection was so great that if you happened to catch one of these infections, your life might be threatened. However, this is still our own history; infectious diseases remain the prominent cause of premature death amongst the populations of the developing world. Almost everyone in the developing world has access to antibiotics, so why has the problem of infection not been eradicated there also? The reasons are many and may have less to do with conventional therapy than we might presuppose.

The earliest texts reveal the impact of infectious diseases; leprosy is described frequently both in the Old and New Testaments. We know that tuberculosis afflicted Neolithic man 7000 years ago; however, despite the terrible impact that these diseases exacted on the population, early man did not understand even the basic rules needed to contain them. Even as late as 300 years ago, when the plague returned to London for the last time in 1665, basic procedures to contain the epidemic were not implemented. Those well enough fled to the country, the remainder took their chances in the city. The dead were rapidly removed but buried in pits because religious belief did not allow them to be burnt. The extended handling of contaminated bodies would only exacerbate the problem.

It was actually during the seventeenth century that the causative agent of many of these diseases was first seen. Antoni van Leeuwenhoek was a Dutch businessman but his hobby was to grind lenses and build microscopes. These prototype instruments comprised a metal plate into which was mounted a single lens, which was held close to the eye. He was the first to describe the single cells that we now know as bacteria but was not able to ascribe a pathogenic role to them. In the eighteenth century, smallpox was the most contagious of the infectious diseases. While practising medicine in the west country of England, Edward Jenner noticed that dairy maids did not later contract smallpox if they had previously contracted the relatively mild disease cowpox. In a totally unethical study, he inoculated an eight-year-old boy with material taken from cowpox pustules and noted that the boy developed cowpox. Several weeks later Jenner inoculated the boy with smallpox, but the disease failed to develop. He had, of course, outlined a procedure for a practical vaccination against smallpox but had he performed this in modern times, he would have faced a lawsuit for unethical malpractice. Although this was an uncontrolled study, it did initiate a vaccination programme that prevented subsequent epidemics until the final eradication of the virus from the human population 20 years ago.

Two interesting developments stemmed from Jenner's work; the first was that immunisation dominated the minds of most medical practitioners, for the following 150 years, as *the* practical solution for infection. We shall see that this restricted the freedom to respond when alternative therapies became apparent. Secondly, the original infection to be controlled by invasive therapy was caused by a virus rather than a bacterium.

Ignaz Semmelweis, born in Hungary in 1818, was a physician who noticed that infection was being carried from the post-mortem and transmitted to women in the maternity wards. Semmelweis demanded that all physicians in his unit wash their hands in an antiseptic solution of chlorinated lime before examining patients. This simple expedient considerably reduced cross-infection, though unfortunately it did not prevent Semmelweis himself dying from puerperal fever in 1865.

It was also in 1865 that, the Edinburgh surgeon Joseph Lister began dipping bandages and ligatures into carbolic acid, and pouring the acid directly into wounds, greatly decreasing the rate of death from gangrene. This was the first use of antiseptics to reduce infection after surgery. He applied Pasteur's theory of germs and he concluded that microorganisms were causing the infection in his patients' wounds. To combat further the effects of infection, Lister also introduced gauze dressings and sterile catgut ligatures.

The greatest of all microbiologists is considered to be Louis Pasteur. He started as a chemist, becoming professor of Chemistry at the University of Lille in 1854. He demonstrated that spoilage of food and drink was caused by microbial contamination and he concluded that infections were also caused by microorganisms. This premise, which was not understood until the middle of the nineteenth century, forms the basis of the control of infection. Until then, it had been assumed that life was spontaneous, exactly as detailed in the book of Genesis. Pasteur had demonstrated that fermentation of sugar to alcohol was a process that required the metabolism of a microorganism, yeast, and that it was not a chemical reaction that yeast simply catalysed. He suggested that many microorganisms were capable of similar fermentation reactions. If microorganisms could use sugar as a medium to grow then perhaps they could also use wounded human tissue; thus he proposed the germ theory of infection. Rather than the spontaneous generation of life, microorganisms are the progeny of other microorganisms; in some manner they could multiply and produce offspring.

Pasteur not only demonstrated that microorganisms caused contamination but he also showed how it could be averted by destroying the causative

Louis Pasteur

microbes, either by heating or by some other means. So although Lister was the first to outline the principle of chemical disinfection, Pasteur showed that heat could sterilise, and although most medical supplies are now sterilised by heating in steam under pressure, known as autoclaving, the heat sterilisation devised by Pasteur, Pasteurisation, is still employed in some parts of the food industry. He championed the need for improved hygiene in food preparation and for greater sanitation, particularly employing sterilisation, to prevent the spread of pathogenic bacteria in hospitals.

Pasteur noted that a bacterium caused anthrax, a disease that afflicted many country peasants and those working in the leather and wool industries. With Jules Joubert, he noticed that the anthrax bacillus could not grow if the first became contaminated with some airborne moulds. They recognised that this finding may have some implications for future therapy. They were, however, more interested in the approach of Jenner; after all, immunisation was the only proven defence against infectious disease. With an attenuated (disabled) version of the causative organism, *Bacillus anthracis*, they demonstrated that inoculation could provide an effective protection against subsequent infection with this bacterium. They had devised the first effective control of a bacterial infection, nearly a hundred years after Jenner. This success restricted Pasteur's strategy to deal with infection; he believed that immunisation was the only way forward. In the late nineteenth century, rabies was a devastating infection in France, and it was particularly prone in children who enjoy close relationships with dogs. In 1882, Pasteur demonstrated that the causative organism of the invariably fatal disease was caused by a microorganism, and by 1885 he used the first vaccine against rabies. The particular success of the rabies vaccine was that it could be administered *after* infection, as long as it was given early enough and certainly before the development of symptoms.

Such overwhelming success for immunisation was to delay interest in chemicals as therapeutic agents. They were to influence another bacteriologist, the German Robert Koch. Koch was not a renowned academic like Pasteur but a country doctor who simply had an interest in infectious disease. His wife bought him a microscope and he started preparing slides of infected tissue. He was especially talented in the understanding of both the optical principles of microscopy and staining techniques. Thus he was able not only to improve the microscope he used but also to optimise the stains he used to highlight the bacteria he was seeking. He was the first actually to see the bacterium that caused anthrax, *Bacillus anthracis*, in 1876. But medicine in Germany was doctrinaire and the revolutionary observations of a country doctor were not taken seriously by the German medical establishment. Virchow, the most eminent of German physicians, did not believe that bacteria were capable of causing disease. Later, Koch became a full-time bacteriologist at the Imperial German Health Office and, while working there in 1881, found the small bacillus that caused tuberculosis. He still could not get his work universally accepted, so on 24 March 1882, he presented his results at the Physiological Society in Berlin. Koch showed the audience how he had developed specific stains that showed the bacillus as a brilliant blue rod, and went on to demonstrate that in every tubercule lesion, the bright blue rods were present. He explained that this organism was very slow growing but live bacteria could be cultured from these lesions. Koch was one of the most objective scientists of his era; he devised rules to define the bacterial involvement in infection. These rules, Koch's Postulates, demand that the bacterium is present in every case of infection, that it is isolated from the infection and purified by culture in laboratory medium, and that the culture is then inoculated into a suitable experimental animal (which was sometimes human and would be considered unethical nowadays) to cause a similar infection. Finally that infection must contain significant numbers of the infecting bacterium. With this meticulous attention to scientific and medical detail, in one lecture Koch convinced the medical establishments of all major European countries that bacteria were the vectors of tuberculosis. His lecture caused great excitement among all present and transcripts were rapidly translated and published in many European capitals. One member of the audience, Paul Ehrlich, was particularly excited; indeed he later said that listening to Koch that evening in the early Spring of 1882 was the most exciting experience of his life.

Paul Ehrlich

Koch and Ehrlich were to remain friends until Koch's death in 1910. Ehrlich was initially fascinated by the staining techniques that Koch had used in order to see the tubercule bacillus under the microscope. This bacterium is one of the most difficult to visualise in microscopy and Ehrlich was convinced that these stains could be made more reliable. Immediately he started experimenting with staining procedures and within months developed the Ziehl-Neelsen stain that we use today. Five years later, Ehrlich used this stain on a sputum sample from one of his own bouts of coughing and diagnosed pulmonary tuberculosis. Fortunately for our story, it was a mild dose and he lived for a further 18 years, during which he produced the significant discoveries of his career. He was closely involved with Koch's work, initially concentrating on the trypanosomes that Koch had identified as the cause of sleeping sickness in Dar es Salaam, in what was then German East Africa. Koch too believed the conventional thinking of the time – that immunisation was the only way to control infectious diseases. Ehrlich was less conventional, an original thinker with a wild imagination who was not afraid to suggest fantastic ideas. He was mesmerised by the effects of dyes and stains on tissues. Conventional contemporary investigation of the cause of death in post-mortem was to examine whole organs, but Ehrlich argued that if microorganisms could be identified by selective staining and examination under the microscope, the organs could be examined under the microscope if they could be sliced thinly enough (microtone) and stained. This technique was the birth of the science of histopathology.

Ehrlich was particularly interested in distinguishing bacteria from human tissue in his histopathological slides. He noticed that some dyes were taken up by some human or animal cells but not by others. There were dyes that were taken up by bacteria but not by human cells. Indeed there was variation amongst bacteria. By 1884 the Danish microbiologist Hans Christian Gram had demonstrated that all bacteria could be divided into one of two groups. By staining the bacteria with crystal violet and iodine, he was able to show that all bacteria took up the dye which is deep purple. Gram then washed the bacteria with alcohol, causing some bacteria to lose the dye, then he counterstained them with basic fuschin which gave them a red colour. The bacteria that did not lose the crystal violet dye on the alcohol wash are now called "gram-positive", and are identified by their deep purple colour under the microscope. Those that lose the stain on the alcohol wash and are highlighted by the red counterstain of the basic fuschin are known as "gram-negative". Ehrlich explained the selective binding of dyes by differences on the surfaces of the cells. The cells of mammals, including man, are surrounded by a membrane whereas all bacterial cells are encased in a rigid cell wall. Ehrlich immediately recognised that the differential take-up of stains might be exploited. Suppose, for instance, that the dye was toxic; if some cells bound the toxic dye then they would be killed, whereas cells that did not bind the dye would remain unaffected. If the dye bound to bacterial cells, but not to mammalian cells, then bacteria could be specifically targeted and selectively killed thus enabling an infection to be eradicated without harming the patient. At the time Ehrlich's hypothesis was visionary and he called this elusive chemical his "magic bullet", but the problem was how to realise the dream. He discussed his idea with Koch, who thought the idea fanciful, but Ehrlich persevered with this notion for the rest of his life. He started by testing one dye after another, examining literally thousands of them. He found that one of them, trypan red, effected a cure in a single mouse which had been infected by the trypanozome that causes sleeping sickness. However, it failed to work in humans with the same disease. Ehrlich realised that his approach was simplistic; it relied almost entirely on luck. After years of research he had just one limited success. Why not alter part of the dye molecule to improve its powers? So he took double nitrogenous dyes and altered them little by little. After each minor alteration, the new molecule was tested. Ehrlich was meticulous at cataloguing his results and he numbered these compounds sequentially. The first compound to have significant success was compound 606, later called salvarsen. In 1909, Ehrlich synthesised arsphenamine, a drug that could destroy the syphilis-causing organism *Treponema pallidum*.

It is difficult to recall the impact that syphilis had at the start of the twentieth century. In many respects, it was viewed in the same manner as AIDS was in the 1980s, except that syphilis was much *more* prevalent. It is estimated that up to 10% of the population of Britain were infected. The disease instilled the same fear that AIDS does in the modern era and this fear probably had as much to do with the lack of sexual promiscuity at the turn of the century as rigid Victorian moralistic teachings. Like AIDS, it was primarily a sexually transmitted disease and there was no effective cure. Also like AIDS, it could be transmitted from mother to foetus, hence the bold statement in the Bible, "The sins of the father.. from one generation to another". It was not quite as deadly as AIDS; only about one-third of those infected went into the fatal tertiary stage, when the bacterium could enter the spinal cord, causing difficulties with sensation and movement, and the brain where it causes devastating effects on rational thinking – many mental asylums of the time were filled with these cases of general paralysis of the insane. Some patients suffered skin and bone damage, and others had damage to the blood vessels around the heart, resulting in heart failure and sometimes requiring surgery. Drastic measures had been used in an attempt to cure it, including exposing the patient to anopheles mosquitoes carrying the malaria parasite, in the hope that the rise in temperature at the onset of malarial fever would kill the rather fragile treponema.

The impact of salvarsen was overwhelming and many syphilitic patients were treated. The side-effects of salvarsen were extremely unpleasant, as it contained an arsenic derivative. These were so severe that some patients died from the therapy; many of these studies were not conducted under the controlled trials that we now use to test pharmaceuticals and it is difficult in retrospect to establish the extent of drug-related deaths or to estimate how many more people would have died if the drug had not been used. Ehrlich did not, however, regard salvarsen as the magic bullet that he had been seeking; it was simply too toxic. He never did find the drug he was looking for and despite being awarded the Nobel Prize for Medicine in 1908, he considered his work to have been a failure. Ehrlich had outlined the principle of selective toxicity and had demonstrated the preferential eradication of cells by chemicals. This was the first example of chemotherapy. He died in 1915 a disappointed man, but we have to regard him as the father of antibacterial therapy.

The German chemical company I.G. Frabenindustrie was an important manufacturer of dyes mainly for the textile industry. At the time that Ehrlich was testing salvarsen in 1910, the company set up the first research laboratory specifically equipped to develop pharmaceuticals. Wilhelm Roehl had worked with Ehrlich on the selective binding of dyes to the trypanosomes that caused sleeping sickness and was certain that the future lay in these selective chemicals. Roehl was appointed as the first research assistant at the experimental laboratory at Eberfeld and was given the freedom to experiment with azo dyes under the directorship of Heinrich Hörlein. With the resources of a major company he was quickly able to repeat Ehrlich's cure of sleeping sickness in a mouse, something that Ehrlich had been unable to do himself. Ehrlich's initial observation had been that dyes could bind to some cells much more readily than to others, which Roehl surmised should not be particularly surprising; after all, dyes were employed in the textile industry because they bound well to natural fibres. Roehl found, however, that it was not the dye that was responsible for the curative properties. The dye trypan red was, in fact, a mixture of chemicals and he found that the active component was a colourless component, the molecule of which chemists could manipulate to improve its activity. The most effective derivative, Germanin, was found in 1916 and was the first effective cure for sleeping sickness.

Gerhard Domagk, a German from the Brandenburg district, was just 23 at the signing of the armistice in November 1918. He had been an infantryman and was wounded in the head in the trenches on the Russian front in 1915. He had been sent back to hospital in Berlin. While he was recuperating, Domagk trained as a medical assistant and was sent back to the Eastern front to work in the battle hospitals. He was horrified by what he saw and recognised that most deaths did not come from direct war injuries but rather from the festering of wounds. This experience made Domagk a life-long pacifist and convinced him that his vocation should be that of a doctor. Within a month of the armistice, Domagk had started his medical studies in the University of Kiel. After graduation, Domagk was appointed as an assistant in the Pathological Institute under the directorship of Walter Gross. Gross had been interested in how the body fought off serious infection and Domagk was directed to observe how the blood dealt with bacterial invasion. Domagk noticed that the injection of mice with *Staphylococcus aureus* initiated a migration of killer cells in the blood that met and then engulfed the invader, known as phagocytosis. He demonstrated that if the animal had first been immunised against the bacterium, the number of killer cells was greatly increased. His most important observation was if the bacterium was disabled or damaged in some way, this phagocystic process went into overdrive and these cells engulfed the debris. These cells were part of the body's defence system to remove foreign matter from the blood. Domagk speculated that if the system was more efficient with damaged bacteria, all that therapy would have to achieve was to injure the infectious bacterium and the body would then readily clear the infection. Hörlein was fascinated by Domagk's observations and recruited him to work in his experimental pathology laboratory at I.G. Farbenindustrie in 1927.

The experimental chemical laboratory was enjoying some success. Roehl had built on his discovery of Germanin to produce an anti-malarial drug, plasmoquine, which was much more potent than the only available therapy, quinine. The drug came to market in 1929, at the very time that Roehl developed a carbuncle on his neck and died at the early age of 48. It was ironic that he should have succumbed to a bacterial infection, one of the

Gerhard Domagk

few organisms to which I.G. Farbenindustrie had not found a potential chemical cure. Domagk was given the freedom to research into whatever he
thought was important and he made the decision to ignore all bacteria
except the one that had killed Roehl, *Streptococcus*. This bacterium was
responsible for a myriad of infections, many of such severity that infection
resulted in death or serious disability. In particular it caused both scarlet and
rheumatic fever. This highly infectious bacterium was able to invade and
damage the heart valves, the kidneys and the liver. Domagk knew that he
would have to set up a large systematic search for a chemical that could
have the suitable selectivity against the pathogen needed for widespread
clinical use. He felt that the limitations of Ehrlich's studies were that the
drugs he devised were not tested on a sufficiently rigorous model, so he
devised his own. He isolated the most virulent strain of *Streptococcus* he
could find, one that had caused rapid, fatal blood-poisoning in a patient from
a nearby hospital. He argued that testing of new chemicals in the test-tube
might give a false impression of their strength because, as he had already
noticed, blood contained cells that could dispose of damaged cells.
Therefore, how much more effective it would be to test his new compounds
in an animal model; this would be a much more sensitive indicator to the
power of these chemicals as he would be measuring not just the ability of the
tested compound to inhibit the growth of the *Streptococcus*, as he would in a
test-tube, but also the capability of the host to capitalise on this damage and
remove the injured bacterial cells. This was a revolutionary concept because
maintaining the *Streptococcus* in an animal model also preserved its virulence; it is a strange fact that when most bacteria are cultured successively in
artificial laboratory media outside an animal host, they can often rapidly lose
their pathogenic characters.

Josef Klarer was the chemist given the task of synthesising the new
compounds, while Domagk injected them into mice that had previously
been infected with the *Streptococcus*. Klarer concentrated his efforts on the
dyes that Roehl had previously examined to cure sleeping sickness. As Roehl
had done before him, Klarer altered these compounds little by little. He provided Domagk with 300 new chemicals to inject into his infected mice, but
none of them were effective. At the end of 1932, Klarer gave Domagk a new
red dye to test. Klarer was pessimistic about its success but Domagk insisted
that it should be tested in his animal model. The dye, which Klarer had
called KL-730, was injected into infected mice under exactly the same protocol as all its predecessors. After the same period of time when all the mice
had died in the previous experiments, Domagk found that all the mice
treated with the drug were very much alive and were showing no symptoms
of infection. Perhaps the mice had not been infected sufficiently with the
Streptococcus or the bacterium had lost its virulence? However, Domagk had
also infected a control set of mice, which were not treated with the dye, and

every single mouse in this group was dead. Domagk and his colleagues immediately took sections of tissue from the surviving mice and examined them for the presence of the round bacterial cells, which formed long chains so characteristic of the *Streptococcus*. Under the microscope, he found the organs of the untreated mice to be saturated with live bacteria but he could find no evidence of any intact bacterial cells in the organs of the treated mice. The drug was called Prontosil rubrum but is generally just referred to as Prontosil. Domagk had been prepared for this discovery because of the testing protocol that he had devised. He was not interested in the results in the test-tube, only in the capability to cure infection in the intact animal. If the failure of Prontosil to show any antibacterial activity during the laboratory testing had been acceded to, then the discovery would never have been made. In fact, a number of antibacterial drugs are relatively ineffective in the laboratory; they have to be metabolised by the host to release an active component.

Domagk tried his new drug against all types of infection but he quickly noticed that it was ineffective against gram-negative bacteria and was more effective against some gram-positive bacteria than others. It was, however, always exquisitely active against the *Streptococcus*. It was readily absorbed in the gut after being taken by mouth, so the drug survived the barrier of stomach acid. It was also freely excreted through the urine; the body could tolerate and did not accumulate it. To improve the solubility, the drug was prepared as a sodium salt and was available for human trials.

These trials started at the Municipal hospital in Wuppertal under the clinical guidance of Dr Philipp Klee. An 18-year-old girl was admitted to the hospital suffering from a severe sore throat, caused by *Streptococcus pyogenes*. Within two days, a severe fever started and the bacterium had caused large abscesses behind her tonsils. These were lanced and the fever subsided. However, the fever returned and she had acute nephritis, a severe condition which results in almost complete blockage of the kidneys and a subsequent failure to produce urine. This condition was invariably fatal. Klee treated the girl with regular intravenous Prontosil injections. Within 24 hours, the patient's temperature was normal and urine was passed freely. The course continued for a further six days and the patient was eventually discharged completely cured.

Prontosil had been discovered in the laboratories of a large industrial company, which was going to be unwilling to disclose its discovery until it was certain that the drug was effective and safe. Too many miraculous cures had been heralded and later found to be too toxic. Prontosil underwent clinical trials at the Wuppertal hospital for two years. Many infections, once fatal, were suddenly curable but they were largely those caused by *Streptococcus* and the related gram-positive bacterium *Staphylococcus*. Nevertheless, the clinical trials demonstrated objectively that Prontosil was an effective treatment for some bacterial infections and Domagk published his results for the first time in 1935.

Prontosil was not readily accepted throughout the world; for some reason Domagk's results were treated with much speculation. Perhaps it was because he was reporting from a chemical company; patent medicines were widespread and it might be that the medical profession considered that this was just yet one more. Some three months after Domagk's initial report, Prontosil was used for the first time to treat a patient in the United States. The woman had bacteriaemia, an infection of the blood caused by *Staphylococcus aureus,* and she was on the point of death. Her family had read of this miracle cure from Germany and insisted that she be treated with Prontosil. Her physicians were openly hostile to the idea but complied. She recovered, although her physicians never attributed her startling improvement to the drug. In England, the drug was first used by Leonard Colebrooke, working at Queen Charlotte's maternity hospital. Semmelweiss had been overwhelmed by rampant infections in recent mothers, and had tried to take preventative measures; however, here in one of the most modern hospitals of the time, nearly one hundred years later and approaching the middle of the twentieth century, the problem was not significantly better. Colebrooke had no weapons against these sudden and often fatal bouts of infection. He had tried salvarsen, Ehrlich's drug, which apart from having extreme side-effects was not very effective and two-thirds of the patients that he treated still died. In 1936, he treated 38 infected mothers with Prontosil and 35 survived. When further groups of patients were treated, the proportion of successful cures increased to 100%. Prontosil could be completely successful if given early enough once symptoms were developing and could still be effective even in the previously premortal stages of infection.

Domagk's own daughter, Hildegaard, pricked herself in the hand and the wound became infected by *Streptococcus.* Her whole arm showed signs of inflammation and she was rushed to hospital. The bacteria entered the blood and the hospital recommended amputation of the inflamed limb, which was conventional treatment at the time for such infections. Domagk, whose faith in Prontosil was absolute, treated his own daughter and she made a complete recovery. It was, however, Prontosil's success in curing the child of another famous father that guaranteed universal acceptance of Domagk's discovery. In late 1936, the son of the President of the United States, also called Franklin D. Roosevelt, had developed severe tonsillitis and the infection was beginning to spread. Eleanor Roosevelt remained by her son's bedside, convinced that death was imminent. As his fever increased, his doctor tried Prontosil as a final desperate measure. FDR junior made a spectacular recovery and the press hailed the miracle drug; the headline of the *New York Times* read "New Control for Infections". When the Roosevelts' doctors were interviewed they were cautious about the role of Prontosil in this most public of successes, and the press picked this up and stated how American physicians were unenthusiastic about the drug but still rushed to use it.

Prontosil was patented by I.G. Farbenindustrie but then, as now, competitors sought ways to "bust" the patent. Jacques and Thérèse Tréfouël, with their colleagues Frédéric Nitti and Daniel Bovet, manufactured Prontosil in their own laboratory in France. They proved that the active component was not the red dye at all; the compound was metabolised by the body's own enzymes and broken down to a colourless constituent. This constituent was active in the test-tube where Prontosil was not. It was identified as sulphanilamide, the first of the sulphonamides. The irony of the French discovery was that sulphanilamide had been manufactured back in 1909 in the University of Vienna but, of course, had never been tested against bacterial infections. Instead it was patented for use in the manufacture of dyes by a chemical company, I.G. Farbenindustrie! By the time the French discovery was published the patent on sulphanilamide had long expired so could be manufactured by anyone who had the resources to synthesise it. Pure sulphanilamide was used in preference to Prontosil, as it was less unpleasant to take. So I.G. Farbenindustrie was unable to profit from its pioneering investment. However, the company did take part in a race to find other sulphonamides that could inhibit bacterial infections that Prontosil and sulphanilamide had been unable to control. Until the world went back to war in 1939, the most wanted prize was a cure for tuberculosis. Sulphanilamide had stopped the spread of tuberculosis in guinea pigs but only when given in the very highest doses. It was not penetrating the protective cell wall of this most elusive of bacteria. Domagk persuaded his chemists to manufacture a myriad of sulphonamide compounds, which he then tested on the tubercule bacillus. At the very time that Hitler annexed Czechoslovakia in March 1939, one of Domagk's sulphonamides, sulphathiazole, was demonstrated to be much more effective than the rest at arresting tuberculosis infection. Domagk has had virtually no recognition for this discovery; when 5% of a continent's population is being slaughtered by war, there is no interest in cures for slow, wasting diseases such as tuberculosis. Indeed the Nazi regime considered it undesirable to seek drugs that saved life when the population was being asked for massive sacrifices. Domagk did eventually present his results, two years after the war had started and at a conference in Vienna. Only delegates from the Axis powers and a few from friendly neutral countries attended, and no enthusiasm for his discovery was reported.

This indifference to Domagk may well have been directed at him personally. Just as Britain and France had declared war, Domagk received a letter from the convenor of the Nobel Committee for Physiology and Medicine. Domagk had been awarded the Nobel prize for 1939 in recognition of his discovery of the antibacterial properties of Prontosil. The Nazi administration had been discredited by the Nobel Committee that bestowed the Peace prize; German dissidents had been awarded the Peace Prize while languishing in concentration camps. Hitler decreed that it was un-German to accept a Nobel prize. Domagk was nervous about accepting this honour and requested permission. The Chancellor of the University of Munster thought that the Peace Prize should be regarded separately from the Science Prizes; after all, the committees were based in different countries. The Science Prizes were decided by committees in Sweden but the Peace Prize was decided upon by a committee in Norway as it was considered that the Swedes were too war-like to deliberate on such a prize. The German authorities remained silent, neither telling Domagk that he could collect the prize nor directing him to refuse it. In November 1939, Domagk was arrested by the Gestapo on no declared charge, though Domagk was always convinced that it was because of the prize. He was released a few days later. He then travelled by train to Berlin at the end of November to deliver a lecture. Domagk was arrested again and, at Gestapo headquarters, he was coerced to write to the Nobel Committee refusing the honour.

Domagk's search through the sulphonamides resulted in Mesudin, a sulphonamide that was active against anaerobic bacteria, in particular the *Clostridium* species responsible for gas gangrene. These infections often occur in limbs and, unless amputated, result in almost certain death. Gas gangrene is a common complication of battle wounds but when Domagk tried to convince army medical corps, no notice was taken. Thus, at the start of the Second World War, Germany possessed powerful new antibacterial agents that could alleviate the gross suffering of the First World War but edict and prejudice delayed their introduction; it cannot be estimated how many hundreds of thousands of lives were lost through this dogma.

These experiences temporarily destroyed Domagk's spirit; he was denied contact with other scientists and he was always in fear of his liberty. He went into a state of depression, exacerbated by his inability to convince the Wehrmacht of the importance of sulphonamides to treat gangrene. The leader of the army surgeons in occupied Brussels was a Dr Wachsmuth, who was opposed to the use of the sulphonamides. He had been persuaded to invite Domagk to Brussels to try out the sulphonamides under field conditions. On the first morning, Domagk was taken to a lecture hall where he met Professor Pfeisseler from Hamburg. It was Pfeisseler's job to recreate the gas gangrene condition in front of the collected audience, mainly comprising army surgeons. He inoculated laboratory bacterial growth medium with earth suspected to contain spores of *Clostridium perfringens*. Open wounds were created in laboratory rats and the earth mixture poured into them. One-half of the rats had the sulphonamide powder sprinkled onto the wound whereas the remainder were left untreated. Within 24 hours, all the untreated rats were dead but all the treated animals survived. The army surgeons were immediately convinced, and the sulphonamide was soon packaged into small containers and became standard army issue for immediate disinfection of battle wounds. Domagk became ecstatic and felt confident enough to renew his investigations into the control of tuberculosis.

Domagk recruited the chemist Behnisch, a man who owed him an enormous personal debt as Domagk had pushed for the use of the experimental Prontosil to treat Behnisch's mother when she contracted bacteriaemia. Behnisch was an enthusiastic disciple and embarked on a painstaking mass of chemical syntheses to find this new, elusive magic bullet. Paul Fildes, an English chemist, had realised that the reason that sulphonamides were active was that they contained a ring structure that emulated *para*-aminobenzoic acid, an essential component for the metabolism of the bacterium. Domagk thought that only two sulphonamides could have any success against tuberculosis: sulphathiazole, which had proved successful against gonorrhoea, and sulphathiodiazole, the drug that he had shown was successful in treating tuberculosis in guinea pigs. However, the ability to penetrate the tubercule bacillus was not related to this part of the molecule, rather it was due to the atoms around the sulphur atom. In the original sulphanilamide, the sulphur comprised a thiazole ring. There was no penetration in tubercule. So Behnisch tried opening the ring. The first compounds he synthesised had no increased activity, and some were found to be very toxic for humans. But Behnisch found that one thiosemicarbazone structure showed the greatest promise.

By now, the company's resources to continue research were beginning to dwindle. Many of the research team were commandeered for other duties and, although the factory at Elbefeld had not been heavily bombed, research into cures for tuberculosis was not considered a major aim in the war effort. I.G. Farbenindustrie was the major manufacturer of sulphonamides and production had to be maintained but the resources of the company were being diverted elsewhere. Ironically, I.G. Farbenindustrie, now the world's greatest producer of life-saving drugs, was also comandered to become a major supplier of a chemical, Xyklon-B, the cyanide-based poison used to implement the final solution in the Nazi extermination camps. Eventually work had to stop at the industrial laboratory as the Allies moved eastwards towards Berlin.

At the end of the war, the British occupied the industrial regions surrounding the Ruhr, in what is now the state of North Rhine-Westphalia. Domagk was allowed to return to work in October 1945, where a British officer who had been seconded from the British chemical company May and Baker attended all strategy meetings. I.G. Farbenindustrie was dissolved by the war commission and a new company emerged, Bayer AG. Bayer had previously been the trade name of the marketing company for pharmaceuticals.

The need for cures for tuberculosis was becoming acute. Labour and concentration camps were opened and the Soviets had occupied all of Eastern Europe. Europe had literally millions of refugees; many were under-nourished and most were herded together. These were ideal conditions for the spread of infection and the incidence of tuberculosis rose steeply.

Behnisch's thiosemicarbazone had considerable activity against the bacillus; if given in sufficient concentration it could eradicate even the most virulent tuberculosis bacteria in guinea pigs. The drug, Conteben, was now, in 1946, ready for clinical trial. The first controlled studies were undertaken in a sanatorium near Münster. The first patient was a volunteer who had suffered from tuberculosis of the skin of the face and had endured 30 years of ulceration. So hideous was her disformity that she had become a permanent recluse in the sanatorium. She was started on a course of Conteben and within three weeks the ulcers started to dissolve and in five months all visible signs of inflammation had vanished. There was now no evidence of the tubercule bacillus, she was completely cured and there was never any evidence of recurrence. Twenty-six patients with tuberculosis of the skin were treated with Conteben. Although not every patient had the same dramatic cure, there was sufficient indication that the drug should undergo larger clinical trials. When Conteben was first used to treat a young woman suffering from meningitis, caused by the spread of the tuber-cule bacillus from the lungs to the meninges, it was administered in high concentrations. The patient died, but before death, analysis of her blood showed that her white blood cell count, and therefore her own immunity, fell to zero. Did the tuberculosis kill her or did the therapy? Our present knowledge of sulphonamides suggests that the patient died directly from the side-effects of the high doses of Conteben, which depresses the bone-marrow production of white blood cells. In order to enter sufficient patients in this trial, Domagk personally travelled around Germany, by horse and buggy because the scarcity of petrol. During this trial Domagk was invited again to Stockholm to receive the Nobel Prize for Medicine and Physiology in December 1947.

A year later, Domagk's old friend and frequent clinical collaborator Philipp Klee had acquired sufficient results to recommend much lower doses of Conteben, thus avoiding the severe side-effects found with the initial meningitis patients. Lowering the dose also reduced minor side-effects such as vomiting and nausea. Between 1947 and 1949, 20,000 patients were treated with Conteben, with remarkable cure rates. As ever, Domagk's newest discovery was largely going unrecognised by the rest of the world. In late 1949, Dr Walsh McDermott travelled to Germany to meet Domagk and returned with a package of Conteben to test. The American press exploited the story and the world was now expecting that tuberculosis was finally vanquished.

The rest of the world had not been idle in the search for cures for tuberculosis. In 1940 the Swedish physician Jorgen Lehmann started following a series of papers written by a friend of his, the American Frederick Berheim, about an observation he had made that if aspirin was added to a culture of *Mycobacterium tuberculosis*, the bacterium took up oxygen much more quickly, so aspirin was enhancing the metabolism of the bacterium. In 1941, Berheim extrapolated that modification of aspirin, salicylic acid, might be able to inhibit the metabolism of the microorganism so he and his colleague, Arthur Sax, manufactured a tri-iodobenzoate. When they challenged the tubercle cultures with this drug, instead of enhancing the metabolism, tri-iodobenzoate inhibited it. Berheim noticed that the new compound did not kill the bacterium; if the cells were transferred to media lacking the compound the bacteria would start to divide again. Berheim could not explain why this chemical should inhibit tubercule. Lehmann noticed the close similarity between the chemical structures of sulphonamides and aspirin, and argued that the active component of these drugs was an amine group; this had been suggested by Fildes' description that sulphonamides emulated para-amino-benzoic acid, and thus Lehmann suggested that a similar chemical group should be added to salicylic acid. In para-amino benzoic acid and sulphonamides, the amine group is located distal, known as para, to the acid component, COOH for para-aminobenzoic acid and SOH for sulphonamides (see figure below). Lehmann would not undertake this himself but he wrote a letter outlining exactly what was required and sent it to the Swedish chemical company, Ferrosan. After some internal deliberation, Ferrosan attempted to manufacture para-amino-salicylic acid. Lehmann was eventually given a few grams to test against tubercule bacilli. Lehmann chose the disabled, non-

Comparison of the structure of para-aminobenzoic acid (left) and its structural analogues sulfamethoxazole (right)

pathogenic BCG strain and started to dilute the para-amino-salicylic acid; it was capable of inhibiting the growth of the bacterium even when diluted one hundred thousand times. When Lehmann reported the results back to Ferrosan, they immediately stepped up production and inoculated their novel compound into rabbits to test the toxicity. They found no serious side-effects in any animals that they tested. It was all very well to try the drug against a non-pathogenic strain, but would para-amino-salicylic acid be active against the most virulent strains of tuberculosis?

At great personal risk, for he had no previous experience working with bacteria as pathogenic as *Mycobacterium tuberculosis,* Lehmann tried the drug in guinea pigs infected with virulent strains. He extended his trials to other infected animals, rabbits and mice, and all were successfully cured. None of these answered the questions as to whether this drug worked in humans and was it thoroughly safe? To test the latter, Lehmann dosed himself both by mouth and by injection. The first clinical trials started 13 weeks after the first manufacture of para-amino-salicylic acid. He treated two children with tuberculosis of the bone and both showed remarkable improvement within weeks, but Lehmann had treated them topically. The tuberculous lesions were purulent on the skin surface so they could be treated by spreading the para-amino-salicylic acid on the surface; the drug did not enter the patient's body.

At the end of October, para-amino-salicylic acid was tested on a patient with pulmonary tuberculosis. A woman who had developed tuberculosis during pregnancy earlier in the year was treated with six grams, which she took by mouth. Within four days, her temperature fell, only to rise again a week later. She was given another course of para-amino-salicylic acid but this time the therapy was more aggressive. Higher doses were given and for a longer period of time. Her temperature fell to normal and her coughing ceased. The other symptoms disappeared but the patient did suffer from nausea; like Conteben, this drug also had side-effects if given in high doses. After further successive courses, the patient was completely cured within six months. No tubercule bacilli could be identified in previously infected tissue. Lehmann would not publish these results; he wanted further clinical evidence before he heralded a wonder cure. This was in marked contrast to his incautious approaches to Ferrosan a year earlier when he had virtually given this revolutionary concept to this chemical company.

Para-amino-salicylic acid had been manufactured for a different purpose during the nineteenth century and Ferrosan was experiencing difficulties in patenting the drug, though it did eventually secure patents for the manufacturing process in several countries. Lehmann published his results for the first time a year later and, by then, nearly 50 patients had been treated for pulmonary tuberculosis. Lehmann first discussed his results at a meeting of the Swedish Medical Society in late 1945. His results were treated with great caution by leading pulmonary physicians; after all, could the remarkable remissions not just be spontaneous? Lehmann stressed the extraordinary coincidence that these remissions *only* followed treatment with para-amino-salicylic acid. The statistical analysis of double-blind clinical trials on antibiotics was still some way off. The phenomenal speed with which the concept of para-amino-salicylic acid had been turned into an active compound and then tested first in animals and then in man was counter-balanced by the delay in informing the world. When the results were finally presented, there were other cures that had been proclaimed for tuberculosis and they have overshadowed Lehmann's brilliant concept. What Lehamann had achieved was quite revolutionary; he had identified the problem and then gone through the intellectual process of designing a drug to inhibit a specific bacterial target. All other antibacterial discoveries up until then had been by serendipity; the discoverers were astute in their recognition that they were observing something out of the ordinary, but they were observing effects of drugs that many had found by happy good fortune, whereas Lehmann actually conceived the drug from the outlet. This was, and still is, a revolutionary concept because, as we shall see, almost all radical antibacterial discoveries have been found by luck. In view of the unique approach that Lehamann had used, it is particularly painful that he did not share the Nobel Prize for Medicine and Physiology in 1952 when it was given for the discovery of anti-tuberculosis drugs. The world failed to recognise a great scientific tactician.

The manufacture of para-amino-salicylic acid was redesigned and Ferrosan could now supply kilogram quantities of the drug; it was distributed to three distinguished chest physicians in Scandinavia. Vallentin, the most eminent of all, had initially been the most sceptical but as the results of miraculous remissions in all types of tuberculosis began to trickle in, he became convinced that they were experiencing a major breakthrough. Vallentin made a cautious statement at the end of a physicians' meeting in Gotenburg in June 1946 stating that significant cure rates had been achieved with para-amino-salicylic acid. The Scandinavian press broke the story and highlighted it as a great Swedish discovery. It was, however, an ill-fated meeting, for many of the delegates did not believe that para-amino-salicylic acid was capable of curing tuberculosis. Many were criticising the drug in private.

Para-amino-salicylic acid also had vocal critics; it was, after all, freely available to only a few physicians. The senior chest physician at Svenshögens, Dr Forgren, attacked para-amino-salicylic acid, stating that it was actually responsible for death in some tuberculosis patients. He based his view largely on two cases in which both patients had been treated with para-amino-salicylic acid initially and then, for different reasons, the therapy was discontinued. At a later point para-amino-salicylic acid was re-administered to both patients and death had soon followed. Dr Forsgren attributed these deaths to the capability of para-amino-salicylic acid to form cavities in the lung and thus aid the tubercule bacillus. In reality, what Dr Forsgren was probably experiencing was the development of resistance to para-amino-salicylic acid; the follow-up treatments were aimed at bacteria that were simply no longer susceptible to the drug. It is a pity that Dr Forsgren's highly subjective attacks overlooked this vital conclusion which would eventually control the manner in which later anti-tuberculose drugs are used even today; it might have brought him honour rather than infamy.

The remarkable results found with analogues of para-amino-benzoic acid and thiosemicarbazones sent many pharmaceutical companies into flurries of research activity. In Bayer's rebuilt laboratories, Domagk had been working on derivatives from thiosemicarbazones, in particular a hydrazone derivative. The chemists had inserted a pyridine ring instead of the benzene ring and the compound, called isonicotininic hydrazide or isoniazid, was 10 times more active against the tubercule bacillus than any other. When Domagk presented his results in 1951 in the United States, it was evident that other pharmaceutical companies had discovered the molecule. In particular, Herbert Fox outlined in three papers at the same conference that the laboratories of Hoffman-La Roche had discovered the same molecule. The molecule had also been synthesised by the Squibb Pharmaceutical Company and was being tested by Walsh McDermott, one of the men who had visited him in Germany in 1949 and had taken Conteben back to the United States.

Hoffman-La Roche and Squibb compared their results directly. The Hoffman-La Roche derivatives of isoniazid had undergone a rigorous test; they had been used to treat nearly a hundred cases of pulmonary tuberculosis. In particular, the drug had been targeted against nearly 50 patients who had tuberculosis in both lungs, and in whom the only prognosis was death. In modern terms, the choice of these patients was quite extraordinary as they had already failed to respond to every other treatment. All patients had a high temperature and X-rays showed the disease was well advanced. But following treatment, every patient was still alive six months later and many were evidently completely cured. Some were well enough to undergo operations to remove the moribund tissue left by the cured tuberculosis lesions, while others were simply discharged back into the communities they had never expected to see again.

These results were considered to come from a chemical manufactured in the United States and it represented a great American success. Domagk was indignant; after all, had he not been the first to declare his results publicly in the United States. In retrospect he must have felt that he had been incautious to proclaim his results without corroborative clinical data, because at least then there would have been no doubt. Actually none of the three pharmaceutical companies invented isoniazid; like many of the other synthetic antibacterial drugs, it had been synthesised some 30 years earlier and none of the three companies was able to patent the compound. Isoniazid was easy to manufacture and could be made readily available for all tuberculosis sufferers. McDermott was worried that this very availability would result in abuse of the drug and, in the end, lead to its failure. He conducted some long-term studies with follow-up on Navajo Indians. There was a high endemic infection rate amongst this ethnic group and he noticed a significant relapse rate after treatment. This relapse rate was particularly worrying because when the patients were treated again with isoniazid they often did not respond to the second course. They were witnessing the emergence of bacterial resistance to isoniazid. During the course of the first treatment the bacteria somehow learnt to overcome the action of the drug; when the second course of therapy was given the bacteria still "remembered" the resistance that they had learnt first time round. McDermott's experiences were mirrored elsewhere wherever isoniazid was used – the tubercule bacillus readily acquired resistance to it.

At the start of the twentieth century, Ehrlich prophesied that chemicals could be synthesised that would inhibit bacterial cells but, because they did not bind to human cells, they would be free of side-effects. Fifty years later this goal was a reality: isoaniazid was a drug that was extremely active against the tubercule bacillus, it was free of major side-effects and could effect remarkable cures; the handicap was the rapid development of resistance and so the era of miracle cures looked to be brief. Isoniazid was not effective against most other bacterial infections and although synthetic compounds would continue to be developed to the present day, deliverance was to come from another source, antibiotics.

2

That's funny

On 10 January 1849, my great-grandmother Ann celebrated her ninth birthday. Nine days later, on 19 January, her uncle James, aged 22, died from cholera. James lived with Ann's parents and five weeks later on 22 February, Ann's baby sister Mary died from the same infection. This was not in some distant corner of the British Empire but in the north-east corner of Ayrshire in the lowlands of Scotland. Examination of the register of Loudoun parish reveals that many people died within a short period of time and this represented a significant community epidemic. The speed of the infection and the fact that young adults, as well as infants and the elderly, died so rapidly suggest that the diagnosis truly was cholera. This was four years before John Snow demonstrated that cholera was an infection that was largely transmitted by contaminated drinking water. Loudoun parish lies along the north side of the valley of the Irvine river. In the early nineteenth century there had been massive expansion of the weaving industry in the valley with a concomitant increase in the working population. Most of the population living in the Irvine valley would have taken their drinking water either from the river or from wells that may have been contaminated by the river. There was no recognisable sewerage policy at that time and it is highly likely that the river was readily contaminated. This would have been exacerbated in the early months of the year because the river has always been prone to flooding during the winter.

Although this infection did not have the same impact as bubonic plague, it did decimate the population of Loudoun parish. In the same parish lived a young man, Hugh Fleming, with his wife and their four children. This family seemed to survive the cholera epidemic intact. Hugh Fleming had leased a farm, Lochfield, from the Earl of Loudoun and his family's salvation may simply have derived from living on a farm on the hills high above the valley. The drinking water here would have been far less prone to contamination. Hugh Fleming's first wife died and on 17 March 1876 he married for the second time, a neighbour's daughter, Grace Morton. He was now 60 and the farm was already being run by his eldest son Hugh. Grace gave him four further children, Grace, John, Alexander and Robert. Alexander was born on 6 August 1881 and after attending the little hillside school, at the age of 12 he went to Kilmarnock Academy. My grandfather remembered him as an intelligent but rather dour scholar.

The weaving industry was rapidly becoming mechanised and work in the valley became scarce. The farms were unproductive and could sustain only one family. With an uncompromising Presbyterian objective, many of the young men from the area, particularly younger sons, left for university. There was stern pressure upon them to study one of the three learned professions, the kirk, law or medicine.

Alexander's elder brothers had already left for London and just short of his fourteenth birthday he joined them. His brother Tom had set up as an oculist in the Marylebone Road and Alexander attended classes. In order to keep himself, he worked in the offices of the American Shipping Line. At the end of the nineteenth century, Britain mobilised 450,000 for the South African War and Alexander enlisted in the London Scottish Regiment. However, he never saw action in South Africa but became a prodigious water polo player. He had just been left a small legacy from his uncle, and with this money, finally settled on following a medical career. Unfortunately this Scot did not have the English Secondary Schools Certificate required to enter medical school in London. In July 1901, he took the exam and came top of all British candidates. He chose St Mary's Paddington as his medical school merely because of the success of their water polo team. He excelled at water polo and rifle shooting as well as being the top student of his year. He had decided on a career in surgery but the bacteriologist Almroth Wright recruited Alexander to his laboratory, to boost the success of his laboratory's rifle team.

During his student years, the young Fleming was exposed to the impotence that clinicians faced with infectious diseases, particularly those following surgery and childbirth. Wright had started an inoculation department at St Mary's hospital in 1902 and this unit was well established by the time Alexander started working for him in 1906. In the 100 years since Edward Jenner's discovery of cowpox vaccination at the end of the eighteenth century, no further progress had been made in combating infectious disease. Wright believed the future lay in vaccination and tried to impose compulsory immunisation on members of the army leaving for the South African war. He stated that "The doctor of the future will be an immunizer." Wright was friend of the immunologist Metchnicoff and was greatly influenced by his discovery of the human body's defence against bacterial disease; the engulfing of bacteria by phagocytes. Immunology was not the only influence on the Inoculation unit. Paul Ehrlich, the preponder of the magic bullet, was a known visitor. However, Wright was not convinced that Ehrlich's views were correct and publicly stated in 1912 that "Chemotherapy of human bacterial infections will never be possible." George Bernard Shaw was also a frequent visitor to the unit and during his discussions with Wright, devised the plot of his play *The Doctor's Dilemma*. Shaw based his character Sir Colenso Ridgeon on Almroth Wright.

Wright is sometimes derided as a pathologist of the old school, a man for whom science was less important than theory. His eloquent dissertations and lively lectures were occasionally based on doctrine rather than fact. In his later years, he was overshadowed by his prodigy and in retrospect some of his ideas appear very naïve. However, his influence on the young Fleming should not be underestimated. Alexander could never have been exposed to such a plethora of ideas without Wright's patriarchial influence. He was, however, an uncompromising researcher and would often disagree with the "Chief". There was a major flaw in Wright's premise that immunisation should be for all: it presupposes that all infectious diseases can be forecast and that all patients will be diligently immunised before any threat of infection. Although Wright would not have been able to forecast the developments of modern medicine, his theory also presumes that all patients will be immunoproficient; in other words they have an intact and fully functional immune system. Fleming was certainly strongly influenced by Wright's theory and was a keen supporter of immunisation but he was less blinkered than Wright. In 1908 he finished his medical studies, obtaining the Gold Medal for that year from the University of London. In the same year, he wrote a thesis entitled *Acute Bacterial Infections* for which he obtained the Cheadle medal. In the discussion of how to combat infectious bacterial disease, this substantial work was not limited by Wright's ideas but rather proposed a multifactorial approach that included surgery, draining of blood-lymph from infected areas, vaccines and raising the patient's own defence system. Although he gave most credence to Wright's theories on immunisation he also raised the role of antiseptics and, more important, the possibility of chemicals to kill bacteria selectively within the body. How he had been influenced by discussions with Paul Ehrlich in the first decade of the century is unknown but, unlike Wright, his mind remained open to other ideas. He was becoming a scientist who could accept observations without prejudice and interpret them dispassionately.

By modern scientific standards, Alexander Fleming might not always be considered entirely objective. In the absence of suitable patients, he would inoculate himself with his staphylococcal vaccine, trying to establish the optimum route of administration to provide the maximum protection. His experiments may not have been tightly controlled but they provided him with a background to accept new ideas.

Alexander Fleming had remained a private in the London Scottish Regiment until five months before the start of the First World War. He rejoined the service, was promoted to captain and joined Wright's field laboratory in Boulogne on the northern French coast. The incalculable carnage of the trenches revealed that Wright's policy of prophylactic immunisation was quite unsuited to dealing with the problems of mechanised warfare. Wright had campaigned to make anti-typhoid immunisation compulsory and it is certain that this policy did save many lives. However, the vast number of open wounds inflicted by explosions as devastating as TNT and the sheer volume of machine-gun bullet wounds produced a deluge of casualties never experienced before. These wounds were usually inflicted in damp trenches and became infected after contact with mud or dirty clothing. In no previous war, save perhaps the American Civil War, were so many wounded soldiers afflicted by septicaemia and gangrene. But this was the first war in which the cause of infectious disease was known. Infection in civil hospitals had been partially checked by the introduction of antiseptics and Florence Nightingale's improved nursing techniques. However, this was infection on a devastating scale. Captain Fleming used his multifactorial approach; he noticed that phagocytosis was increased in these wounds and he suggested that removing necrotic (dying) tissue to improve phagocyte penetration would aid the body's own defence. There was significant pressure to use topical antiseptics and large quantities were used in the dressing of wounds. However, the staff of the pathology laboratory in Boulogne observed that they were virtually useless. These chemicals are not selective in their action and may kill some of the human cells they are trying to protect. In minor surface infection, the sacrifice of some cells may not be harmful and the presence of the antiseptic may prevent further infection by bacteria, but amongst the war wounds experienced in the trenches, very few were merely surface infections. Captain Fleming made a simulated war wound model in a glass vessel and demonstrated that in a deep-rooted infection, however many times the wound was washed with antiseptic, the cause of the infection was never removed, lying deep in the tissues. The antiseptic merely served to kill the surrounding tissues and impeded the body's own phagocytes reaching the infection. He realised that no chemical available at that time had the power to cure deep bacterial infections. Therefore, under Wright's guidance, he performed a series of experiments to stimulate phagocytosis. They demonstrated that strong saline solutions promoted the phagocytic process. When Wright returned to England and propounded the use of saline solutions as phagocyte promoters rather than antiseptics, he was derided by Sir William Watson Cheyne, and enthusiastic supporters of Lister who saw Wright's comments as a direct attack on their work. This put Alexander Fleming in an ironic situation, because here he was in 1916, arguing against the use of chemicals in the treatment of infectious disease. Indeed, when he gave the Hunterian Lecture in 1919, entitled "The Action of Physical and Physiological Antiseptics upon a Septic Wound", he still proposed stimulation of the body's own defence system over the use of chemicals.

In the early 1920s, Alexander Fleming returned to his laboratory at St Mary's and started performing experiments on the antibacterial properties of body secretions. He was particularly interested in tears as he had demonstrated that they contain a substance that rapidly kill bacteria. In retrospect such a powerful bacterial killing agent might seem an obvious component of tears – how else did the mucosal surfaces of the eye overcome infection? – but in the 1920s this was not apparent. He demonstrated the power of tears by taking a culture of bacteria, adding one drop of tear and within seconds the murky culture cleared; the bacteria had been broke down or lysed. The tear contained a much more powerful component than any of the antiseptics that they had been experimenting with during the war. The substance was named *lysozyme*, an enzyme that lyses bacteria. With the help of his student, Fleming demonstrated that almost all bodily secretions contained this enzyme, and he concluded that it was protecting the vulnerable mucosal surfaces against infection. They also found lysozyme within the body, in the blood and especially in the white blood cells.

They turned their attention to other animals and flowers. Some animals and many flowers contained copious quantities of lysozyme but the most prolific was egg-white. It was highly effective at killing bacteria even in very low doses. Its purpose is to protect the egg yolk from infection as the yolk is a highly nutritious medium for bacteria as well for ourselves.

The discovery of the antibacterial properties of lysozyme was Fleming's first major scientific discovery. He tried to capitalise on it and demonstrate that lysozyme could be used as an effective treatment option. Employing a technique now much favoured by microbiologists, Fleming took a Petri dish containing nutrient agar. He cut a hole in the agar at the centre of the dish and filled it with the teardrop lysozyme. He prepared pure cultures of bacteria known to be pathogenic to man and streaked these bacteria along the

Sir Alexander Fleming

surface of the agar from the hole to the outer circumference of the dish. The lysozyme should diffuse out of the hole and through the agar. If it was effective against the bacteria it should prevent the bacteria growing at some distance from the centre. Many non-pathogenic bacteria tested by this technique were found to be susceptible to the action of lysozyme; however, when pathogenic bacteria were tested, they were not affected and usually grew right up to the hole containing the lysozyme. Although disappointed, Fleming could explain his results by suggesting that pathogenic bacteria cause infection because *they* possess the capability to overcome the body's own defences, including lysozyme.

Despite this setback Fleming believed that it could be possible to use lysozyme in large quantities. He was unable to concentrate it and his inability to purify antibacterial products was to obstruct his research throughout his career. Had he been a proficient chemist, or employed a dedicated biochemist, he would undoubtedly be hailed as the most important microbiologist of all time, Pasteur, Koch and Ehrlich not withstanding. However, purification was not possible so Fleming looked for a more profitable source of lysozyme and returned to egg-white. With egg-white lysozyme, he could show that almost all pathogenic bacteria were, at least, partially susceptible to its action. A genus of bacteria, *Enterococcus*, is known to colonise the intestine of some people and is known to have some pathogenic properties. Fleming persuaded a patient who was excreting large quantities of these bacteria to swallow the whites of four eggs – the level of enterococci fell to normal levels. These studies were extended to several volunteer patients with the same syndrome and the enterococcal levels in all volunteer fell. However, this was an uncontrolled study and was not conclusive. He continued these experiments only with patients whose lives were threatened and had limited success. However, the medical community at the time either ignored his publications or were actively hostile. In 1927, he made one further discovery with lysozyme, the importance of which has been totally ignored by medical science. When he treated *Staphylococcus aureus* or enterococci with increasing concentrations of lysozyme, some colonies were able to continue to grow. In fact, some of the original bacterial population had mutated; they had "learnt" how to grow in the presence of the enzyme and they were now resistant to the action of lysozyme. The presence of lysoszyme would kill all those bacterial cells that had not mutated, so now only the mutated cells could grow. He also found that the mutation improved the bacteria's susceptibility to the natural killing action of blood. The acquisition of reduced susceptibility to lysozyme was also accompanied by increased virulence and greater protection against the body's defences. These results were the first example of bacterial resistance to a specific antibacterial drug; they were also a harbinger for the future success of antibiotics but these results were totally ignored by fellow scientists. Scientific research should be the impassioned and methodic test of hypothesis; however, like many other human pursuits, it is prone to fads and fashion. Insignificant work may be heralded simply because of the vibrant personality of the proponent whereas important observations may be dismissed simply because their premise is unfashionable at the time. It may also be that the scientist or the research group is unpopular and their publications are simply ignored. I have already mentioned that Fleming was not a flamboyant personality and this rather dour Scot was an outsider within the London medical community.

Throughout the early 1920s, Fleming continued looking at antiseptics. He repeatedly faced the same problems that he experienced in the First World War; most agents he tested were very capable at killing bacteria in a relatively clean environment. If there was any body tissue present, the efficiency of the antiseptic was considerably compromised. The chemicals bind to receptors on many types of living cells, human as well as bacterial. Besides reducing the capability of the chemicals, these results revealed the true flaw of antiseptics – they simply were not selective. Fleming showed that they destroyed the phagocytes but they also destroyed other human tissue. Injecting these chemicals to treat infections would result in not only a weak end-product but also a drug that had devastating side-effects. Fleming recognised this but his work on antiseptics and lysozyme had prepared him for his greatest discovery, penicillin.

It is popular nowadays to denigrate the contribution that Fleming made to the discovery of penicillin. He is often portrayed as a rather naïve scientist who, totally by chance, made an observation that was to change medical treatment for ever. This is far from the truth for he had already worked for 20 years on potential treatments for bacterial infections; he *knew* what to look for. The story of how he observed the potential of penicillin is now a legend. He was clearing old Petri dishes away from a previous experiment while talking to his assistant Merlin Price. Fleming was not one to throw anything out without examining it carefully first. His attention was drawn to an old agar plate containing colonies of *Staphylococcus aureus* that had been contaminated by a mould.

In a famous understatement he remarked unemotionally to Pryce, "That's funny." Although contamination of Petri dishes is quite common if they are left, he had noticed that unusually the colonies around the mould had lost their colour and become translucent. This was very similar to the effects he had observed previously with lysozyme. He surmised that, like lysozyme, there was a substance produced by the mould that diffused through the agar and killed the bacteria. It had to be a small compound otherwise it would not diffuse through the agar, so here was a chemical, produced by a mould, that could kill bacteria. With great forethought, he took some of the mould for further subculture.

He subcultured the mould onto agar plates and obtained colonies identical to the one he had observed on the original plate. He placed a number of pathogenic bacteria on similar plates in close proximity to the mould; many of those tested, including staphylococci, streptococci and enterococci, were inhibited, and so also were the bacteria responsible for diptheria and tetanus. However, the mould did not inhibit salmonella, including the species responsible for typhoid. Fleming cultured the mould in large vessels in a nutrient broth, and knowing that the active chemical exuded from the mould itself, he removed the mould and concentrated on the remaining broth, which had turned a brilliant yellow. This compound was already in the hands of a prepared mind, for Fleming carried out exactly the same experiments on this extract that he had performed with lysozyme some six years earlier; he prepared a gutter in the agar and placed the extract in it. The bacteria to be tested were streaked outwards from the edge of the gutter. He found that the extract was at least as powerful as the original mould. He started to dilute to find its strength. Even when he diluted a thousand times, it was still well capable of inhibiting staphylococci. It was much more powerful than lysozyme, but was it safe? The mould was identified as *Penicillum notatum*, and he called the compound penicillin. The mould was a fungus closely related to that responsible for producing the blue streaks in stilton and other "blue" cheeses. Human ingestion of these cheeses has not been associated with significant side-effects, so perhaps, unlike the antiseptics, this compound was not toxic. Fleming recognised that he was witnessing a phenomenon where one organism was targeting the destruction of others around it to improve its own chances of survival. This was called antibiosis by the nineteenth-century French scientist Vuillemin, so Fleming classified penicillin as an antibiotic.

It is clear that Fleming was not the first to observe the effect of penicillin but he was the first to recognise its potential. Hyssop, mentioned in the Old Testament, is believed to be the first example of the healing properties of penicillin. In 1871, Lister had noticed that the presence of a fungus that he identified as *Penicillium glaucum*, placed on the surface of a nutrient medium, rendered that medium clear for subsequent culture of bacteria. He did not pursue this research. In Manchester, England, William Roberts noted that the growth of fungi could prevent the growth of bacteria, and vice versa. He noted specifically that *Penicillium glaucum* was immune to bacterial infection. Louis Pasteur noticed that if animals were injected with *Bacillus anthracis* they rapidly developed the symptoms of anthrax; however, if some non-pathogenic bacteria were injected at the same time, the animal was protected from anthrax. Pasteur recognised that there may be some therapeutic value in this observation but could identify no way to exploit it. There are also a number of anecdotal observations in the eighteenth and nineteenth centuries that some patients with severe gastrointestinal disorders had significant alleviation of symptoms after eating blue cheese.

Although Fleming recognised the medical significance of penicillin, his problems started when he tried to obtain pure extracts. He demonstrated that it was unstable in high (alkaline) pH and if heated. He also showed that the extract was not toxic; large quantities could be injected into animals with no apparent side-effects. However, penicillin was still prepared in a nutrient broth from which the producing *Pencillium notatatum* was removed. The penicillin extract still contained a large number of unknown proteins, peptides and other contaminating material. This had been Fleming's shortcoming with the development of lysozyme, and similar difficulties were found with penicillin. The laboratory still did not have a dedicated chemist and, had they had one, they would have almost certainly have been able to purify pure antibiotic. Two recently qualified physicians were given this extremely difficult task and they nearly succeeded. Unfortunately neither had any experience in chemical purification to fall back on and learnt all their techniques from books. They tried to concentrate penicillin by reducing the volume of the broth by applying a vacuum. They could not boil away the liquid because pencillin was already known to be heat-sensitive. They managed to concentrate the antibiotic about 50-fold but were left with a brown glutinous mass which was quite unsuitable for therapy. This penicillin also rapidly lost its potency, even if kept in the refrigerator.

Fleming was not personally involved in these attempts; he concentrated on the microbiological observations. He resolved that he should present his initial work, which he did at the Medical Research Club on 13 February 1929. His presentation was dry and dull, he stuck rigidly to the facts and failed to fire any enthusiasm for the potential of his results. He repeated the failure that he had previously had when he presented his results with lysozyme. He could not excite his colleagues. No questions or queries were raised after his paper and this deeply wounded him, so he decided to publish his results in the *British Journal of Experimental Pathology*. In this paper, he suggested that penicillin had the potential to be effective in therapy. This conclusion angered Wright, who still believed that stimulation of the immune system was the *only* method to overcome bacterial infection. If Fleming had convinced Wright of the importance of his results and had Wright publicly supported his protégé, then a greater importance would have been placed on his results. After all, Wright was the antithesis of Fleming, a flamboyant extrovert, and he would have been a far more persuasive ambassador for penicillin.

The inability to purify penicillin did not deter Fleming. He tried his most concentrated extracts on open wounds with a little success but many of these trials, by their very nature, were uncontrolled and it was difficult to interpret them as a decisive success for penicillin. Thus penicillin went into limbo for nearly 10 years. Fleming never lost his interest in penicillin. In 1935, Domagk was invited to give a lecture at the Royal Society of Medicine in London. It was the first time that Prontosil had been revealed and Fleming was in the audience. While Domagk outlined the potential of his new antibacterial drug, Fleming remarked "Yes, but penicillin can do better than that." He noted that the concentrations of Prontosil that Domagk was using were high and the effects were less dramatic than his own. His frustration became even more acute.

In 1935, the Chair of Pathology at the Sir William Dunn School of the University of Oxford was filled by Howard Florey, a 37-year-old Australian whose particular interest was chemistry. In the years leading up to his appointment, he had been a Rhodes scholar and this had first introduced him to Oxford. A subsequent Rutherford scholarship enabled him to work in a number of different laboratories in the United States. Florey had been particularly interested in Fleming's rather forgotten papers on lysozyme and on his appointment he assigned two chemists to purify it, which they first achieved in 1937. One of these chemists was Edward Abraham. Florey was looking for experts in the new field of biochemistry and he appointed Ernst Chain, a Jewish refugee from Nazi Germany, who had studied at the University of Berlin. Florey persuaded Chain to study the bacteriolytic activity of lysozyme. Chain showed that lysozyme was an enzyme and he postulated that, as it was specifically toxic to bacteria, it must have a unique target in bacterial cells. In order to establish what this target might be, Chain searched through all the literature on substances reported as known to have antibacterial activity. Here he found Fleming's paper on penicillin from 1929. It seemed to Chain that the new substance that Fleming was describing was much more promising than lysozyme. He was particularly impressed by its greater potency and almost complete lack of toxicity. Fleming had been surprisingly free in giving cultures of *Penicillum notatum* to any scientist who requested it and the department in Oxford had obtained a culture long before Florey ever became interested in antibacterial drugs. Florey was a competent biochemist but a novice mycologist; he had significant difficulties in growing the fungus. When he could persuade it to grow, it did not always produce penicillin. He concluded that penicillin was highly unstable. He thought it might be an enzyme like lysozyme, as enzymes often lose activity when they are concentrated; the impurities that surround them are also concentrated. To overcome this problem, enzymes are often freeze-dried. The solution is frozen and then a vacuum is applied. The water is removed by sublimation; the vacuum extracts it as a vapour because it never melts. The impurities are inactive in both the frozen and solid states. Freeze-drying takes the solution through the former state and concentrates the constituents directly to the latter state. After freeze-drying, Chain obtained a much more concentrated penicillin preparation with much greater antibacterial activity. However, he had also succeeded in purifying many of the impurities. Chemists usually remove the chemical that they wish to purify by extraction, and place it into a solution into which it dissolves easily but into which the impurities dissolve poorly. Fleming's results suggested that penicillin could dissolve in alcohol, which Chain thought paradoxical because he believed that penicillin was an enzyme and enzymes are destroyed in alcohol. Nevertheless, Chain tried to extract penicillin with pure ethyl alcohol and he failed. Then, with stubborn persistence that defied his better judgement, Chain tried further extractions with other alcohols. Reason would advise that this was a wild-goose chase; preconceptions about enzymes would suggest that these experiments were doomed. He found that methyl alcohol could extract and thus purify penicillin. In the presence of concentrated methyl alcohol, penicillin again became unstable but Chain found that with the simple expedient of diluting the alcohol in water, the stability improved dramatically. When they freeze-dried the extract again they had pure penicillin. Fleming had used alcohol extraction and failed to reach the final goal simply because he had evaporated the extract in the liquid phase rather than frozen-solid stage.

Chain had mice injected with 30 milligrams of pure penicillin, equivalent to injecting humans with more than 50 grams, and there were no observed side-effects at all. Florey urged Chain to set up a larger-scale purification process. Chain was not a microbiologist and was not experienced in assaying the activity of antibacterial substances. He recruited Norman Heatley and together they set up an assay. When fully purified, penicillin was one million times more active than Fleming's extract had been. After further toxicity studies on different laboratory animals, they started infecting mice with streptococci; a control group (25) were infected but not subsequently treated whereas the other group (25) were injected with pure penicillin every three hours after infection. This was the first controlled experiment to be performed on an antibiotic. All the untreated mice were dead within 16 hours, while all but one of the treated mice survived for the whole duration of the experiment. This was one of the most miraculous experiments ever performed by man; Chain, Florey and Heatley sent their results to the *Lancet*. Fleming read this article but we don't know how he truly felt about this finding. He travelled unannounced to Oxford and Chain was astonished to see him as he assumed that Fleming was dead because he had not continued publishing his studies. Fleming formally congratulated Chain and thanked him for realising the potential of his discovery; on his return he said of the Oxford team, "They have turned out to be the successful chemists I should have liked to have with me in 1929."

Penicillin had to be tested on human subjects so the Oxford team tried to improve the yield; Heatley was given the task of extraction and Chain and Abraham concentrated on its purification. They managed to precipitate about 500 units of the barium salt; a unit is the smallest quantity that will produce a zone of sensitivity of 2.5 cm diameter with *Staphylococcus aureus* and is still the unit used for measuring this antibiotic but no other. It had be demonstrated on a life-threatening case but the problem was how to administer it. It is unstable in acid, so it would destroyed by the stomach acid if given by mouth. It was also thought to be excreted rapidly from the animal host through the kidneys. How this information was discovered is not documented – it may have been extrapolated from the data obtained with mice; however, pharmacological information obtained on rodents is notoriously unreliable if applied to humans. The mice had been treated successfully by injections at three-hour intervals, so this was a continuous infusion. This is now recognised to be the optimum administration of this type of antibiotic but these pioneers would not have known this.

The pivotal case was an Oxford policeman dying of septicaemia; he had a scratch at the corner of his mouth. This had become infected by *Staphylococcus aureus* which had invaded the bloodstream. There were abscesses throughout his body and, on 12 February 1941, he was just 24 hours from death. Two hundred milligrams were injected initially followed by 100 mg every three hours. The improvement was dramatic; the patient regained his appetite and his temperature fell. The supply of penicillin was insufficient to continue treatment and the injections had to stop. As soon as they ceased, the infection returned and, without any defence, the patient died a month later. Although this appeared very encouraging for penicillin, the result was not conclusive. The medical team had also given the policeman blood transfusions and perhaps these had been responsible for his improvement. If they were to prove that penicillin was an unequivocal treatment, they had to treat patients with sufficient quantities of antibiotic in the absence of other procedures, including blood transfusions. When Heatley had extracted sufficient penicillin, the team treated three more seriously ill patients; two effected a complete cure and the third died from a cause unrelated to the infection.

Florey's role was to try to persuade the emerging pharmaceutical industry to take over the manufacture of penicillin. Despite the clinical successes, there was a reluctance to become involved. Most of these companies were either chemical companies more suited to the manufacture of industrial constituents than medicines or were undergoing a metamorphosis from peddling patent medicines. Antibiotics, which many were then ignoring, were eventually going to transform some of these companies into the most successful in the world. At the time, Great Britain stood alone against the Nazi threat and had a war economy, resources were scarce and venture capital was not available for such speculative ventures. The essential problem was cost-effectiveness; in order to prepare sufficient penicillin to treat a single case, thousands of litres of culture had to be extracted and the antibiotic purified. Florey realised that he would have to approach the contacts he had made in the United States when he was there as a Rutherford Scholar. At some personal risk, he and Heatley travelled to Lisbon and took the clipper to New York. They carried cultures of *Penicillium notatum* with them. They were on a mission that they believed was essential for the war effort; like Fleming, they had made no effort to patent their discovery. Like the trawl through the chemical companies, Florey and Heatley tried to interest many laboratories, but none were interested. Only when they arrived at Peoria did they find any interest. The fermentation laboratory had been set up to deal with an unusually high accumulation of corn steep liquor, a by-product of the manufacture of starch from maize. They had already demonstrated that they could make gluconic acid with cultures of *Pencillium chrysogeneum*. Culture of *Pencillium notatum* immediately increased the yield of penicillin 20-fold, and changing the carbon source from glucose to lactose increased the yield further. Heatley stayed in Peoria to try to improve the yield further; the *Penicillium notatum* strain was descended directly from the one that Fleming had subcultured from that first observed agar plate. The yield could only be improved by changing the chemical constituents of fermentation; perhaps another culture might be more productive. The Americans had now entered the war and the army was stationed almost over the whole globe. The group in Peoria requested that the accompanying medical corps obtain fungus cultures from as many sites as possible, so that they could test their potential as producers of antibiotic substances; perhaps one of them would produce more penicillin than *Penicillium notatum*. This is a technique still used by some pharmaceutical companies but, in this case, the answer was found in Peoria itself. A cantaloupe melon, bought in Peoria, had gone mouldy; this had been caused by *Penillium chrysogenium*, a fungus closely related to Fleming's. This culture increased the output of penicillin but the yield was further improved radically by genetic mutation.

Florey visited many chemical companies both in the United States and Canada and, when he returned to Oxford, two companies had promised to manufacture some penicillin for clinical trials in England. While waiting for these false pledges, the Oxford team tried to boost their own extraction yields, keeping aside their small stock for any clinical emergencies. Some of their penicillin was used to treat pilots who had been badly burnt and some was sent to the army in Egypt where it was used successfully to treated badly wounded patients.

Fleming was the first to use penicillin against meningitis. This disease, still a scourge, can kill within hours after the bacterium enters the meniges, the cavity surrounding the brain. A friend of his, a man of 52, was dying in St Mary's hospital. He had all the symptoms of meningitis but no bacterium could be found in the cerebrospinal fluid. Fleming isolated small numbers of streptococci. He telephoned Florey who placed his entire penicillin stock at Fleming's disposal. As a desperate measure, Fleming injected into the spinal fluid. This was a bold procedure because the Oxford team had performed this procedure on a cat and the cat had died shortly afterwards. However,

Fleming's friend did not and he made a complete recovery. *The Times* picked this up in a leading article on 27 August 1942, declaring the virtues of this elusive drug and challenging the government to invest in its production. Even this article did not reveal the contribution that either Fleming or the Oxford team had made to the discovery. This was left to Almroth Wright who, in a letter to *The Times* on 28 August 1942, acknowledged Fleming's original discovery and its importance to medical practice. This public recognition of Fleming's discovery by his immediate superior was rather belated but ironically it was, perhaps, the one statement that alerted the public to Fleming's contribution. So penicillin was used in an increasing number of infections but, as in the First World War, the most devastating infections were those experienced by infantry after shell attacks. The western allies were beginning to see large numbers of these wounds in the African campaign. This was their first major ground offensive against the Axis powers and the number of wounds contaminated with earth and dirty clothing was extreme. It was further exacerbated by flies laying their eggs in open wounds. So bad was this problem that Florey flew to North Africa to demonstrate the optimum use of penicillin: first test the susceptibility of the bacterium to the antibiotic and then, if sensitive, treat with an adequate dose. Much penicillin was wasted on wounds caused by gram-negative bacteria against which penicillin had no activity; using the drug against bacteria inherently insensitive was an iniquitous waste of resources. Large-scale production was slow both in the United States and in England and it was not until 1944 that the production of penicillin in the United States reached a level that could meet the demand for it.

In 1945, it was announced that Fleming, Florey and Chain would receive the highest scientific accolade, the Nobel Prize for Medicine. The rules of the prize state that there can be only three co-recipients; it was instigated when most research was performed by individuals, rarely by a team. Florey was revolutionary in the way he had set up the Oxford team but the sheer number in this team meant that many would go unrecognised. Would we have had penicillin without Heatley or Abraham?

There were other antibiotic pioneers, who have received far less attention, but their contribution to the use of antibiotics was crucial. Selman Waksman had been born in Russia in 1888 but emigrated with his family to the United States while still a child. He had trained at the Rutgers Agricultural College in New Jersey and then studied for his Masters degree at the University of Washington. While in Washington, Waksman started isolating and identifying new microorganisms in soil samples. This was virgin territory. The microbiology of soil was poorly understood and virtually every microorganism that Waksman identified was until then unknown. One organism that had caught his attention was very unusual; it did not appear to be a bacterium but it was not identical to the larger-celled fungi either. This was

Selman Waksman
(Comes from a Web site excerpted with permission from The Foundation for
Microbology)

an impossible organism to classify; he concluded that it was halfway
between the two and he called it *Actinomyces griseus*. We are now able,
with our vastly more sophisticated detection techniques, to classify this
species strictly as a bacterium but to Waksman its origins remained mysteri-
ous. Waksman completed his Masters degree and, after a short spell away on
the west coast of America to study for his PhD, he returned to Rutgers
College as a soil scientist specialising in microbiology. Waksman was an
aggressive researcher and gathered a team of scientists, each with his own
speciality, and this team continued ploughing through endless soil samples
searching for ever more exotic and elusive miniature forms of life.

René Dubos was a Frenchman born in 1901 but always harboured a
passion to emigrate to the United States. He had graduated in agronomy from
Paris and travelled to the United States in 1924 to seek his scientific fortune
studying with the team of soil microbiologists at Rutgers College, who were
now gaining a world reputation in classification. René Dubos is perhaps the
most underestimated of all the antibiotic pioneers. His start at Rutgers was
fairly inauspicious. Waksman wanted him simply to devise techniques to
count the number of microorganisms in a soil sample. Dubos found this
routine and tedious work; he was a primarily a theorist who liked to plan
research strategy meticulously, not perform the duties of a laboratory assistant. Dubos managed to abandon the project and moved on to examine

humus, as his interest was in decomposition. Humus was full of the debris of death, both of microorganisms and the larger multicellular plants and animals. In this potentially nutritious environment, microorganisms proliferated, but how do they access the proteins and carbohydrates of dying cells? Dubos argued that bacteria must have enzymes that break down the cellulose of plant cells, forcing them to relinquish their nutritional rewards. He persuaded Waksman to let him study cellulose breakdown for a PhD project. In 1926, he had his first success: he found a soil bacterium that could break down cellulose and actually grow on the breakdown products. Dubos presented his results on cellulose in 1927 at a conference on soil science in Washington and was excited by the enthusiastic response. He knew that he had established a crucial concept in the balance of life and he believed that it had wide implications. He also felt constrained by the narrow perspective that the agricultural surroundings of Rutgers offered and he knew he had to move on. He applied for a National Research Council Fellowship, but on the letter of rejection was a suggestion that he should consult Alexis Carrel, a French surgeon at the Rockerfeller Institute, who might be able to help find the position that he was searching for.

When Dubos arrived at the Rockerfeller Institute in New York, Carrel had no vacancies for him but, after their interview, invited him to lunch. They lunched with the head of the microbiology laboratory, Oswald Avery. Avery was interested in Dubos' work on cellulose breakdown and they talked long after the meal had finished. Avery recognised the similarity of his own work on clinical bacteria and was sure that Dubos could exploit his expertise and find a way to break the defences of pathogenic bacteria.

Four months after their meeting in April 1927, Avery offered Dubos a fellowship. Dubos considered this as the turning point of his life; it is difficult to tell because he left the patronage of Waksman, a man who was still to make the most important discovery of his life, to work for Avery, under whose guidance he performed scientific miracles. Dubos was a biochemist; he had had no training either in microbiology or in medical sciences so his appointment to Avery's microbiology laboratory was unlikely. He does, however, seem simply to have interested Avery but this enigmatic appointment was to reap an enormous harvest. Avery was an unusual scientist who believed that thought and discussion were much more important than rushing to carry out experiments, an opinion that Dubos adhered to rigorously. It can be speculated that these two spent hours discussing the profound role played by bacteria in the competitive game of evolution. Dubos saw the scavenging bacteria of the soil competing with each other for the spoils of death while Avery was transfixed by the challenge that bacteria made to human life itself.

Avery was interested in the pneumococcus, the cause of severe and rapid lung infection which could often prove fatal, particularly in the elderly and very young. However, it was invariably fatal if that infection spread to blood. Avery had found that the pneumococcus was surrounded by a capsule, a polysacharride which formed a tough outer layer, and he concluded that this capsule was the reason that the bacteriun was pathogenic; the phagocytes could not engulf it. Pneumococcal bacteria lacking the capsule were very susceptible to phagocytosis whereas the production of this polysacharride coat made the bacterium impenetrable. Avery used exactly the same capsulate bacterium in the 1940s to demonstrate that the carrier of genetic material was DNA.

Dubos acknowledged that the capsule was the key to success. Destroy the capsule and the bacterium becomes sufficiently vulnerable for the body's own defences to take over. Dubos speculated that, as he was seeking a compound that specifically consumed the polysaccharide, he should examine environments where polysaccharides are broken down regularly. He thought that he should examine soil, a medium with which he had become so familiar, for the presence of bacteria that could destroy the pneuococcus. He set up a series of innovative experiments; he made an extract of the capsule and placed soil samples in it. If any of the bacteria in the sample could grow, he surmised that they must be breaking up the capsule and that this was the only energy source available to them. He tried many different bacterial types but met with no success. Perhaps the conditions were wrong, so he tried changing the conditions of growth, altering the temperature and pH; perhaps the microorganisms were wrong. Dubos knew that there must be microorganisms that broke down these capsules; if there were not, surely the world would be weighed down with old capsulate bacteria. He must be looking at the wrong microorganisms.

When he had been at Rutgers, there had been a bog which he remembered contained gelatinous material very similar to the capsule of pneumococcus. He contacted Waksman and asked him to go to New Bruswick and collect samples from different parts of the bog. With this new material in his experiments, he had almost immediate success; the capsule of the pneumococcus was being digested and some of the microorganisms from the bog were proliferating. He needed to find the microorganism that had the greatest capacity to thrive on the capsule, as this would be the most likely candidate to produce the magic bullet against pneumococci. He achieved this by selective subculture, arguing that the most successful organism would be the one that divided the most quickly; in the end it would outgrow all its competitors.

Dubos' experiments did lead to one microorganism growing on the capsule and he isolated this final victor. He grew the organism up in artificial culture and, after removal of the bacterial cells, was left with an extract. He immediately embarked on a trial in mice, first infecting them with pneumococci and then treating half of them with his extract. Untreated mice died rapidly whereas those that had been given the extract survived. Avery joined

Dubos and enhanced the level of analysis, outlining experiments to demonstrate when the extract should be given and observing under the microscope
exactly what was happening to the hapless pneumoccocal cells. As Dubos
had predicted, the characteristic diplococci were systematically stripped of
their coat and abandoned to eventual destruction by the phagocytes. Dubos
and Avery were ready to start human trials but they suddenly heard of the
startling discovery from Germany; Prontosil had beaten them to the final goal
and this *Streptococcus* was vanquished. Both men were devastated, as this
work had become their life's mission. They lacked the tranquil self-belief that
Fleming possessed and saw no purpose in continuing. In fact, they had,
through careful reasoning, devised a scientific strategy that would prove to
be a powerful tool in the search for new antibiotics.

Dubos decided to turn his attention to the *Staphylococcus*. This bacterium, unlike pneumococci, was not always responsive to Prontosil.
Staphylococcus aureus has no outer capsule; it is pathogenic because of the
myriad of toxins that it is able to excrete to ease its passage through the body.
Dubos had no capsule to extract so he modified the experiment and used the
staphylococcal cells themselves, substituting them as the selective agent for
his soil bacteria. By painful and laborious selection, one organism outgrew
the rest. It was a gram-positive spore-forming bacterium belonging to the
Bacillus genus, and he called it *Bacillus brevis* 3/4. Avery and Dubos were
keen to discover the nature of their success, and they quickly identified the
active components. They were to find that *Bacillus brevis* was producing a
cocktail of antibiotics; the first to be identified they called tyrothricin. It was
extremely active against pneumococci and staphylococci, much more so
than Prontosil. It was also stable in solution, which was in stark contrast to
penicillin at that time. Surely they had the first true antibiotic, and certainly
the first to be found by a systematic search. Dubos presented his results at the
Third International Congress of Microbiology in New York in September
1939. Domagk had been invited but had been unable to go as Germany
plunged into war. Fleming was there and was frustrated by the failings of his
own discovery now that Dubos had shown that antibiotics could work.

All that Dubos had to do was purify the active component of his
extract and then they would have a revolutionary pharmaceutical. Dubos
sought help from a chemist in Avery's laboratory, Rollin Hotchkiss. Hotchkiss
was a willing collaborator and found the prospect of developing a new drug
much more interesting than his own esoteric research. They started by
growing large quantities of *Bacillus brevis* and then removing the active
extract from gallons of culture media. The active extract obtained was about
200 grams of impenetrable brown treacle. As Fleming had found, the chemist
can try differential removal of chemicals based on solubility. The extract was
almost totally insoluble in water so they had to try to remove the active component with organic solvents. This was not easy because addition of many

organic solvents seemed to make it even less penetrable. They used hot organic solvents, particularly ether, and they noted with amusement that their techniques were considered to be a dangerous fire risk. As the tyrothricin was removed from the brown milieu, it seemed to show greater and greater relative activity. Then they found an astonishing property: when they performed one of their organic solvent extraction procedures, the tyrothricin activity seemed to separate into two; they were not dealing with one chemical but *two*. Hotchkiss completed the purification and one morning presented Dubos with a pure chemical, which when examined under the microscope was composed of thin straight clear rods. These crystals were very effective at killing bacteria in laboratory experiments but as soon as they tried to cure bacterial infections in animals, they hit a major problem because it was toxic. It did not have a greater action against bacteria, and it seemed to demolish all cells, animal as readily as bacterial. They called the compound tyrocidin, but it had proved to be a blind alley. The other chemical was much more promising; they completed its purification to produce boat-shaped crystals. When they tested the chemical they found that it was active only against gram-positive bacteria and had no action against gram-negative. This was still only November 1939, penicillin had yet to be stabilised and the only drug for bacterial infections was Prontosil, which was also active only against gram-positive bacteria. Dubos and Hotchkiss' new drug would target the same bacteria as Prontosil but it seemed much more active and anyway Prontosil did not always work. They called the chemical Gramicidin because of its preference for gram-positive bacteria. Hotchkiss and Dubos patented their discovery; they were not interested in exploiting their discovery for personal gain, rather they wished to prevent parsimonious pharmaceutical companies from patenting the drug for themselves, so the patent was transferred to the Rockerfeller Institute for one dollar. It seems incredible nowadays that two academics should hand over their rights to their employer for no reward. Their aims were altruistic; they were interested in helping mankind. The Rockerfeller Institute had benevolently supported their whims and provided financial guarantee to their research, a rare occurrence in the modern University structure.

The newspapers naturally discovered the story, and the discovery was hailed as a gigantic leap forward in the mastery of infectious disease. Pharmaceutical companies queued up to start development programmes to identify the chemical structures of tyrocidin as well as gramicidin. These companies also tested the two drugs exhaustively in animals and confirmed that tyrocidin was too toxic but they also showed almost catastrophically that gramicidin was also too toxic; it lysed the membranes of animals' cells but in retrospect this might not be too surprising as it killed bacteria by lysing their membranes. It could never be given for any treatment in the body, whether administered by mouth or by injection. It was, however, suitable for topical use so it could be used to treat infections on the body surface, skin infections, wounds and cuts. It could even be given to treat infection of the nose. This was not a true magic bullet; it was more akin to the antiseptics and would appear destined to have the same role in medicine. In fact, it was much more important than that; the world was already back at war and battle wounds again became the major infection problem in the world. Gramicidin was extensively used in the curing of infections in open wounds and, when soaked in bandages, prevented the establishment of infections. Florey and Chain recognised the discovery by Dubos as the impetus that stirred them to explore the role of penicillin. Penicillin could be given systemically, was much safer and when introduced to treat battle wounds soon relegated gramicidin to a second-division antibiotic. The most important discovery of Dubos was not the drugs he found but the systematic method he had devised to find them. While he was working so intensively on gramicidin René Dubos' wife, Marie Louise, was dying of tuberculosis. Her death in 1942 altered his life; his inspiration had gone and he was never to reach the heights that he previously reached. He moved to Harvard to escape the memories of New York City, but he had to work on infections directly relevant to the new Pacific war. In 1944, the benevolent Rockerfeller Institute invited him back, and provided him with a completely equipped laboratory to work on what had become his single passion, to find a cure for the infection that had taken his wife away from him.

Dubos' first mentor, Selman Waksman, had overseen an experiment at Rutgers that looked at the survival of tubercule bacilli in soil specimens. This was an interesting experiment because *Mycobacterium tuberculosis* looked like a bacterium that used man as its natural host, perhaps even its only host. However, these experiments demonstrated that the bacillus could live well in soil samples, indeed it proliferated in this environment. Could it be that this bacterium was not primarily a human pathogen but had derived from the soil? If it was, other bacteria would have to compete with it and they would probably release antibiotic substances to kill the tubercule bacillus. Surprising they found that if other bacteria were added to the soil, the tubercule bacillus grew more quickly. Was the tubercule bacillus producing antibiotics against these competing bacteria? A much more fascinating observation was that if fungi or fresh manure were used instead, it was the tubercule bacillus that died rapidly. One of the competing microorganisms was producing a killer chemical that could control tubercule; if nature could achieve this, why not man? However, like so many discoveries about the antibiotic properties of natural products, no gain was made from this one either. Perhaps Dubos was the catalyst; if he had still been working with Waksman, this line of experimentation would never have been completed. Frank Ryan speculates in his book that Waksman was still working in an agricultural college; he had no direct interest in clinical infectious diseases.

Unlike Dubos, he may not have had any direct contact with tuberculosis. He did not appear to have a mission to find a cure for infectious diseases; he was more interested in the effect one microorganism might have on the survival of another, but this could be *any* microorganism. He was interested to discover what the killer chemicals were and how some bacteria proliferated in the soil by killing others surrounding it.

Ryan suggests that it was Dubos' ideas that converted Waksman to think of using his concept to control infectious disease. The views of students or even ex-students are often not considered seriously by the prejudiced minds of their mentors; however, the questioning of the established view is the essence of new discovery and Waksman's conversion to the line of thought of his protégé was to give him a Nobel prize and the world a stunning gift but to leave his student unrecognised.

Waksman's conversion to Dubos' thinking was as sudden as Paul's on the road to Damascus; it was as though the mist cleared and the way ahead suddenly became clear. He rushed into his laboratory and told his team to stop everything they were currently working on; that very day, all of them were redirected to look for new antibiotic substances. Boyd Woodruff was a chemist and a recent addition to the Rutgers' team. He, like Waksmann, was not medically qualified nor had he worked with medical bacteria; however, they sat down to target which medical bacteria needed a cure. At the time, all the discovered antibiotics were active against gram-positive bacteria but many infections were caused by gram-negative bacteria, and patients infected by these bacteria could not benefit from the amazing discoveries they were reading about in their newspapers. They decided to "cure" gram-negative infections. They had Dubos' system of progressive selection to find the one substance that provided the best survival advantage but they wanted to show specific activity against one class of bacteria. They prepared soil samples with live microorganisms and added them to the bottom of a Petri dish. On top of the soil, they poured a warm molten agar solution containing a suspension of the pathogenic gram-negative bacteria that they hoped to conquer. As the agar cooled down, it set to form a jelly-like layer over the soil sample. The plates were placed in an incubator, usually at 37° Celsius, the temperature of the human body at which most pathogenic bacteria grow at the optimum rate. Under normal conditions the pathogenic bacteria would multiply and, within 24 hours, the growth would be visible as a cloudiness throughout the agar. The theory was that if any microorganism in the soil were producing antibiotic substances, they would diffuse away from the producing microorganism though the agar and prevent the growth of the pathogenic bacterium. This would produce a result not unlike Fleming's original observation of penicillin killing the surrounding *Staphylococcus aureus*. More importantly, they needed to isolate and identify the microorganism producing the antibiotic substances. This was easy from their technique; the pre-

vention of growth was seen as a clear circle, and at its centre was the producing microorganism. It could be removed, purified from other microorganisms and identified.

They found two antibiotic-producing microorganisms. One came from
a strain of *Pseudomonas aeruginosa,* a gram-negative bacterium found in
many types of soil. This bacterium could also be a human pathogen, particularly in wounds, and was one of those bacteria for which they were hoping
to find a cure. The second was a gram-positive bacterium, an *Actinomyces,*
one of those enigmatic microorganisms that Waksmann had worked on at the
very start of his research career. They called the antibiotic substance
Actinomycin after the microorganism that produced it; however, like the
other antibiotics generated by soil bacteria, when they injected it into
animals, it was toxic and killed mice at the concentration needed to kill bacteria. It was at this stage that they attracted the attention of a pharmaceutical
company, Merck, Sharpe and Dohme. This pharmaceutical company
patented the actinomycin and tried to determine its chemical structure. If
they could achieve this then just maybe they could find a less toxic version.

Waksmann's research expanded, examining all actinomyces that the
team could process. They found one that had a very powerful capability to
kill bacteria. The producer was a light-blue organism classified as
Streptomyces lavendulae. The active chemical was called streptothricin.
Unlike gramicidin and actinomycin, it was readily soluble in water and,
unlike penicillin, it was stable. It was active against all types of bacteria,
killing both gram-positive and gram-negative with equal rapidity. The
Rutgers' team tried limited experiments; it was able to cure cattle of
Brucellosis, caused by *Brucella abortis,* a serious infection because of its
transmission to man. It is manifested by retention of the placenta and the
production of weak young. Infected animals also had difficulty in breeding,
infertility and abortion. All the cattle treated seemed to have miraculous
cures; this was the first time this infection had ever been cured and there
appeared to be no side-effects. With this immense boost, streptothricin was
passed on to Merck for extensive trials. Woodruf went to Merck's laboratories
to help with the animal studies. Suddenly the trials were stopped; mice
treated with streptothyricin were dying from kidney failure. Merck dared not
proceed further with this chemical; it was, like the other soil-derived antibiotics, too toxic. They were so near to producing a viable drug, for the compound was being prepared for human studies. However, they were already
looking at new substances and this disappointment would soon be forgotten.

Albert Schatz joined Waksman in June 1943, as he wanted to study for
a PhD in soil microbiology. Waksman had still not recovered from the blow
of his two failed antibiotics. He would listen to any bizarre theory to control
bacteria and though he probably still believed the future was in chemicals,
he saw this as an elusive dream. He saw the prize as a cure for tuberculosis,

but the two miracle drugs sulphonamides and penicillin were useless. There were some unconfirmed reports that modifications of the sulphonamides might be useful but Waksman was convinced that the answer lay in the soil; after all, he had performed those original experiments that showed that there was something in soil that killed the tubercule bacillus. The best drug he had discovered had been streptothrycin and he thought that actinomyces must be the source of this magic bullet. In Schatz, Waksman recruited not a systemic scientist but an intuitive investigator, for Schatz was given literally thousands of actinomyces to examine; he liked the look of some more than others and chose them for further study. He would set up a screen to evaluate every colony that attracted him; he would cross-streak the colonies with the bacteria that he wanted to test activity against. He would take a loop of the colonies and spread them across the surface of the agar, then at right angles he would take his test bacterium and run that across the plate. If the actinomyces were producing an antibiotic substance, then it would diffuse through the agar away from the actinomyces and inhibit the test bacterium. The test bacterium simply looked as if was unable to grow in that part of the plate. He had no systematic reason for choosing the samples that provided him with new actinomyces, so it is likely that he retrod many paths. He did however keep every strain that he ever tested – it might be useful one day.

Very rapidly he did find actinoyces well capable of inhibiting a variety of pathogenic bacteria, but he could be choosy in which ones he identified for yet more sophisticated tests because some actinomyces produced only a little inhibition whereas others inhibited bacteria some distance from the producing actinomyces streak; he made a fairly simple correlation that the further the inhibition the more active the chemical produced. This correlation is not exactly true; smaller compounds will diffuse through the agar faster than larger ones, so greater inhibitory power might just mean that the agent is smaller.

The activity of the most prolific producers was quantified; Schatz prepared larger quantities of the actinomyces and extracted the antibiotic. He tested their activity by preparing lawns of test bacteria; molten agar containing the test bacterium was poured into a Petri dish and allowed to set. With a sterile glass or metal hollow cylinder, a hole is cut into the agar and the "plug" removed. This provides a well to insert the antibiotic substance. This would diffuse away from the hole and, if the test bacterium was sensitive, would produce a circular zone clear of growth whereas the rest of the plate would be opaque with the vigorous activity of the uninhibited bacterium. The larger the zone, the more active the antibiotic, and a figure could be placed on this as the diameter of zone of inhibition could be measured. At two in the afternoon of 19 October 1943, Schatz found what he was looking for. He found two actinomyces that had very similar and visually stunning characteristics. Schatz was captivated by two actinomyces that produce green/grey colonies. They had come from different sources, one from a specimen taken from a chicken and the other from heavily manured soil. What particularly attracted Schatz' attention was not just the ability of both colonies to destroy the pathogenic bacteria in the cross-streak but the massive extent by which it achieved this. The agar plate was almost completely clear. The zone of inhibition was also totally free of any bacteria so the inhibition seemed absolute. Such a convincing result had never been seen before.

Schatz soon found that the capability of producing organisms to manufacture the antibiotic substance was dependent on its nutrition source, so altering what was added to the growth medium in which they grew the actinomyces enabled them to produce maximum amounts. Their microbiological experiments showed that they had stumbled across a very versatile antibiotic; it was able to inhibit the gram-positive bacteria already under the control of penicillin and sulphonamides but it was also capable of inhibiting the more gram-negative bacteria, which were still unaffected by these new discoveries. Waksman knew that this discovery would succeed only if the substance was not toxic, so he immediately ordered some animal tests. *Salmonella* species are gram-negative bacteria and are common causes of infection in chickens. Waksman decided to infect chick embryos with *Salmonella gallinarum* and then treat them with the new antibiotic. They made a small hole in the shell and added a lethal dose of *Salmonella*; half were subsequently treated with the antibiotic. All those that were treated hatched while the untreated birds died in the egg.

They reclassified the producing actinomyces as *Streptomyces griseus* and thus created the *Streptomyces* genus, a group of soil microorganisms that have been, by far, the most abundant source of clinically useful antibiotics. They named the antibiotic streptomycin after the genus name with which they had now ascribed the producer. Schatz went on to test streptomycin against the tubercule bacillus. He repeated the technique he had used against the other pathogenic bacteria but this time the experiments took much longer; the tubercule bacillus grows very slowly and the results of the experiments would not be known for weeks. When the bacteria had been incubated for some weeks, Schatz was not surprised by what he found; he intuitively knew the result namely that, streptomycin inhibited tuberculosis. The excitement of this result was not allowed to eclipse the necessity to purify streptomycin; inability to purify penicillin had ruined the initial discovery. Schatz performed this himself and found to his delight that the substance was highly soluble in water and was stable in solution. He was able to prepare it in solid form, which would keep and then dissolve readily in water, retaining its high activity.

This discovery was eventually to be passed on to William Feldman, an ex-patriot Scot who had specialised in veterinary science. He worked with

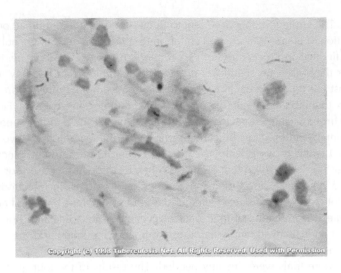

Tuberculosis bacillus

Corwin Hinshaw at the Mayo clinic evaluating new "cures" for tuberculosis by testing them out in experimental animals infected with the bacillus. Much of their time had been spent evaluating derivatives of the sulphonamides. At the time of streptomycin's discovery, they were evaluating Promin, a drug that was active enough to retard the course of infection in human volunteers, but the problem of toxicity, so often associated with sulphonamides, was very evident. The red blood cells were lysed with the treatment and the trial had to be curtailed. They were provided with a variant of this sulphone, called promizole, which had similar anti-bacterial activity to Promin but was considerably less toxic. Feldman and Hinshaw were in the process of setting up a major clinical trial with Promizale when Feldman made a speculative visit to Rutgers to ask Waksman if he had any new antibacterial substances that might be developed into suitable treatments. Feldman was unaware of the research on streptomycin and with the distrust of a scientist, Waksman was not going to enlighten him; however, Waksman was interested in Feldman's view that virulent tubercule bacilli should be tested whenever new substances were discovered. Feldman thought that virulent tuberculosis was the only scourge that really justified a cure and that less serious infections would fall under control once the tuberculosis was conquered. He supplied Waksman with some virulent strains of *Mycobacterium tuberculosis* and Waksman tentatively agreed to send Feldman some of his most promising compounds to test in animals and perhaps humans. Waksman had no facilities to test his compounds extensively in animals and was not able to test in man.

The virulent strains were sent to Rutgers and Feldman still had only the intention of please Waksman to collaborate. Waksman and Schatz published their findings in *Proceedings of Experimental Biology and Medicine*. In their paper they made no significant mention that streptomycin had any activity against the tubercule bacillus; instead they concentrated much more on its ability to inhibit the previously untouchable gram-negative bacteria. Feldman and Hinshaw were contacted by Waksman, inviting them to test streptoycin in animals and humans. By early Spring 1944, the team at the Mayo Clinic had 10 grams to work with. They divided it into four and then infected 12 guinea pigs with a virulent strain of tuberculosis. They chose four of the guinea pigs and started them on a series of streptomycin injections. The antibiotic stock was depleted after 55 days but some of the untreated animals had died. Examination of their organs revealed serious infection but all the treated animals lived, and when these animals were killed and their organs examined, there were no bacteria and little cellular damage.

More trials were urgently needed. The results of the four guinea pigs were striking; but before human trials could begin more guinea pigs must be tested. Schatz simply could not keep up the production of streptomycin so Waksman, Feldman and Hinshaw approached the pharmaceutical giant Merck and Co. Merck had conflicting interests, as they were heavily committed to the manufacture of penicillin. The penicillin was required for battle wounds against which tuberculosis took second place. However, in a bold assurance, the owner of the company, George Merck, personally promised that they would develop a production process that outstripped anything that Schatz could manage as long as Feldman and Hinshaw tested the drug clinically. In a much larger guinea pig study, many animals were infected and the course of the disease was allowed to proceed for seven weeks. Examination of some animals revealed that they were close to death. The animals were then started on a systematic course of streptocycin treatment. All the treated animals survived and they seem to have suffered neither from any long-term effects of the infections nor from side-effects of the treatment itself.

The time now came for human trials. It was first tried *ad hoc* on a humanitarian basis in single, critically ill patients for whom death seemed certain. The first was an old man who had meningitis. Although there was insufficient streptomycin to cure him and the trialists had absolutely no idea what dose should be given, they noticed some improvement in the condition before the patient died. In the second infected patient, the bacteria were established in the blood, kidneys and liver. This patient, who had had high fever and had been in severe pain, made an apparently outstanding recovery; his fever disappeared and he looked set to survive. Unfortunately a thrombosis, a consequence of his confinement, suddenly killed him; however, biopsy of his organs revealed that the infection had been controlled and they appeared to be healing. A woman with pulmonary tuberculosis was the first to be treated successfully. Still the correct dosages had not been worked out and the patient was treated with 100 milligrams over a six-month period. Eventually her temperature fell and all signs of infection disappeared from her lungs. She was still coughing and the sputum still contained active tubercule bacilli but she was recovering. What was the streptomycin doing to the bacteria? Were they reducing the virulence of it? No explanation was offered. The patient was discharged and made a lasting recovery. Further clinical trials confirmed that tuberculosis had finally succumbed.

This was the golden age of antibiotic discovery and opportunity for any self-respecting microbiologist to have a go at trying to find the next miracle drug. Everyone was reading about the systematic approaches, particularly those of Dubos and Waksman; surely there were other antibiotics to be found. The major pharmaceutical companies started their own antibiotic searches. G. Duggar, a scientist working for the Lederle Research laboratories of the American Cyanamid company, was examining soil samples that had been taken from a Missouri farmyard and found an antibiotic substance that was active against both gram-positive and gram-negative bacteria. It was produced by another member of the *Streptomyces* genus, which was golden yellow in colour, so they named the producer *Streptomyces aureofaciens*. They called the antibiotic aureomycin but today we call it chlorotetracycline, the first of the tetracyclines. The antibiotic had remarkably low toxicity and was active against almost all bacteria and even some other microorganisms, such as rickettsia. Rickettsia are a group of small, often disease-causing, rod-shaped bacteria with two genera, *Rickettsia* and *Coxiella*. At one time they were believed to be midway between the larger viruses and smaller bacteria in size but their structure reveals that these submicroscopic organisms are really true bacteria. However, they differ from most other bacteria in that they are parasites that exist only within other cells and are dependent on a host, usually a bloodsucking insect, for part of their life processes. The antibiotic was manufactured in the United States and was slow to come to Europe because of problems with currency.

In 1947, soil from a mulched field in Venezuela was being examined by Bartz, a botanist, who worked in conjunction with a group of researchers from the Parke, Davis and Co. pharmaceutical company. They too isolated a member of the *Streptomyces* genus which was producing an extremely active antibiotic substance. It was similar to the *Streptomyces lavendulae* that produced streptothricin and was named *Streptomyces venzuelae* after its country of origin. A research team from the University of Illinois were examining compost soil from a farm in Urbana, USA and found a very similar organism which produced an antibiotic of almost identical properties. The antibiotic was purified from the culture fluid and chemists identified its structure. They called it D-(-)-threo-2-dichloroactetamido-1-p-nitrophenyl-1:3-propanediol or chloramphenicol for short. It was an unusual compound because it was a derivative of nitrobenzene and dichloroacetic acid; a similar product had not been seen before in Nature so it seemed that this compound had evolved solely for the purpose of attacking competing microorganisms. Parke, Davis and Co. patented the compound and marketed it as Chloromycetin, a trade name it retains to this day. Like tetracycline, it had a broad spectrum, inhibiting both gram-negative and gram-positive bacteria.

This rush for new antibiotics continued around the world. Vancomycin was obtained from an extract from *Streptomyces orentalis*. This was found to bind specifically to the disaccharide peptides that made up the cell wall. This is a fairly toxic drug and for a long time was ignored in favour of the myriad of more active and safer antibiotics. However, as the spectre of resistance emerged and the more active antibiotics became useless, vancomycin was revived and we now rely on it to save lives in some of our most severe infections. The search continued and similar detection techniques applied to an endless stream of soil and sewerage samples found adriamycin, amphoteracin, bleomycin, cycloserine, erythromycin, gentamicin, kanamycin, lincomycin, neomycin, nystatin, oleandomycin, paramomycin, rifampicin, spectinomycin and viomycin just from actinomyces largely from species the *Streptomyces* genus. Some of the antibiotic producers were true bacteria, including *Bacillus subtilis* which produced Bacitracin and *Bacillus polymyxa* which produced Polymyxin. The late 1940s and early 1950s revealed that the world was full of natural substances that could cure even the most serious bacterial disease. The *Streptomyces* genus had been particularly fertile; these soil organisms, which must be fairly aggressive in their own environment, provided almost all the antibiotics that were needed to stem bacterial infection.

The golden age of discovering new natural antibiotics was soon to end, for two reasons. The first was exhaustion of the supply. The enthusiastic search for new drugs continued, of course, but soon there were no new antibiotics. Many new substances were found to be identical to antibiotics that had been discovered and patented years earlier. Many more, however, were just too toxic for human use and as there were many much safer drugs, there was no need to develop them further; just carry on looking for more. The second reason was that bacteria were learning how to survive in the presence of the antibiotics, and were becoming resistant. At the end of the Second World War, approximately 50% of *Staphylococcus aureus* strains were already resistant to penicillin. The march of resistance was relentless, to the point that further drugs might be needed, but none were found. It is particularly interesting to note that by 1954, we had already found almost all the true antibiotics that would ever be discovered; in 15 years all the natural defences had been revealed. If resistance continued a crisis would result, with no new antibiotics to overcome bacteria that had become resistant to the older drugs. A new approach would be required.

In 1945, Professor G. Brotzu had noticed that the coastal waters around his native Sardinia had reduced numbers of bacteria when they were fed by sewerage outlets. He therefore sampled a number of these outlets and isolated some of the microorganisms present. He obtained a fungus that had a very odd-shaped colony; it appeared similar to the lobes of the brain and the genus was named *Cephalosporium acremonium,* literally head-forming spores. Brotzu knew what he had, a fungus like *Penicillum notatum* that was able to produce an active substance that could kill bacteria, but unlike penicillin it was active against gram-negative as well as gram-positive bacteria. Brotzu was, however, a microbiologist and not a chemist. He had read how much difficulty Fleming had had in trying to purify penicillin so decided not to try. He thus decided to send the strain to Oxford, where Edward Abraham, who had worked on the purification of penicillin, identified that the *Cephalosporium* produced two antibiotics which he called cephalosporin C and cephalosporin G. The latter was very weak but cephalosporin C had sufficient activity for development as an antibiotic. The original drug of the cephalosporins was not successful as an antibiotic because there were now many other antibiotics which were much more active. The structure of cephalosporin C was elucidated and identified as a derivative of 7-aminocephalosporanic acid. This was to form the building block of a completely new group of antibiotics, known as the semi-synthetic drugs, which were to be vital in the eventual battle with resistance.

3

Development of modern antibiotics – man's mastery over infection

The late 1950s until the mid-1980s were considered to be the most productive era of antibiotics, but this was not because there was a plethora of new compound types. All the main natural antibiotic groups had been discovered by the mid-1950s and in 1961, the nucleus of nalidixic acid, a 4-quinolone, was identified. This was the last novel antibacterial chemical nucleus ever to be discovered which means that there have been no innovative antibiotics for 37 years, new drugs have been variations of old chemical structures. This should not suggest that these new drugs have been ineffective, because they have been, but rather, if the role of these drugs was to combat resistance,

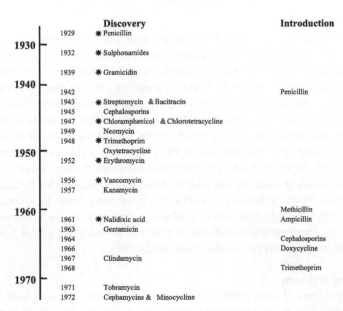

		Discovery	Introduction
	1929	✳ Penicillin	
1930			
	1932	✳ Sulphonamides	
	1939	✳ Gramicidin	
1940			
	1942		Penicillin
	1943	✳ Streptomycin & Bacitracin	
	1945	Cephalosporins	
	1947	✳ Chloramphenicol & Chlorotetracycline	
	1949	Neomycin	
	1948	✳ Trimethoprim	
		Oxytetracycline	
1950			
	1952	✳ Erythromycin	
	1956	✳ Vancomycin	
	1957	Kanamycin	
1960			Methicillin
	1961	✳ Nalidixic acid	Ampicillin
	1963	Gentamicin	
	1964		Cephalosporins
	1966		Doxycycline
	1967	Clindamycin	
	1968		Trimethoprim
1970			
	1971	Tobramycin	
	1972	Cephamycins & Minocycline	

The Golden Age of antibiotics

they would have a more difficult task because they were closely related to drugs that had been responsible for resistance in the first place. The antibiotics that had been discovered by the start of the 1960s were modified essentially for a series of different reasons.

IMPROVED CHARACTERISTICS

The first antibiotics were injectable because they were unable to survive the acid conditions in the stomach, oral drugs have to pass this first natural defence against unwanted or dangerous ingested chemicals. To overcome this barrier, chemicals either have to be made inherently resistant to destruction by acid or have functional groups added to form an ester. After the ester group has carried the antibiotic through the extreme conditions of the stomach, it is then cleaved, often by enzymes in the host, to release the pure antibiotic. The second essential property is that the antibiotic is readily absorbed in the bowel so no residual antibiotic reaches the bacteria-rich region in the large intestine. This may cause devastation of the commensal faecal flora and lead to side-effects, such as diarrhoea. It also provides a fertile breeding ground for resistance. In commercial terms, the advantages of an oral antibiotic are immense and many manufacturers would strive to make oral versions of their most promising compounds. The ability to give an antibiotic by mouth allows the drug to used widely in the community and that is where most of the money in antibiotic sales is likely to be made.

Selective toxicity

The essential property that differentiates antibiotics from antiseptics is that they are sufficiently selective to allow their use within the body, rather than just on the surface. This selective toxicity is not absolute but it is quantifiable as we shall see in chapter 6. There is, therefore, always room for improvements and developing drug licensing regulations seek improvements in the comparative safety of antibiotics. The drug must inhibit the target bacteria at lower concentrations, usually much lower, than those concentrations that produce toxic effects in humans. Some antibiotics can be given in very high doses without toxic effects, e.g. penicillins, but others may produce serious toxicity at levels that are not much above those required for treatment of infection. Many alterations to antibiotics have been made to improve this selectivity; though with some antibiotics this is virtually impossible to achieve. The most selective antibiotics tend to be those that inhibit a process in bacteria that does not exist in mammalian cells.

Dosing regimens

The half-lives of early antibiotics were short, perhaps only one hour so the antibiotic had to be administered many times per day in order to maintain sufficient concentrations. Injectable versions are generally administered in hospitals and the regular dosing of patients becomes very expensive in both money and nursing staff. Oral antibiotics are taken without supervision in the community, this often and perhaps usually causes problems with patient compliance. If the dosing intervals of drugs could be increased then patients are more likely to remember to take the drugs, particularly if they have to take them just once a day. Thus modern modifications to antibiotics often try to incorporate much longer half-lives as part of the properties of the new compound. This might be as long as 33 hours which means that the patient needs to take the therapy just once a day in order to maintain sufficient drug concentrations.

To be or not to be – the spectrum of antibiotics
When I started as a Medical Microbiologist, I was often told that there could only be a single bacterium responsible for the infection. This follows the original Koch's postulates but bacterial infection is often not as simplistic as this anachronistic view might suggest. So the traditional microbiologist will try and identify a single pathogen as the cause of particular infection. If a single pathogen is the cause, then an antibiotic that would just tackle that bacterium, and leave all others unaffected, would be preferred. If antibiotics could be administered in this way, it would cause the least disruption to the body. It is also the epitome of Ehrlich's dream of a magic bullet. It was also the dream of some antibiotic manufacturers but this naïve perception of infection is increasingly outmoded, particularly in infections in the immuno-suppressed. In many cases a broad-spectrum cover will be desirable if the pathogenic bacterium has not yet been identified and therapy has to be started urgently or the risk of super-infection is high. Antibiotics are often described as broad- or narrow-spectrum, according to number of different bacteria that they can inhibit.

Bacterial death – is it essential?
Antibiotics may or may not kill the bacteria. If they do they are called bactericidal, if they merely inhibit replication of the bacteria which remain viable and may start to grow when the concentration of drug falls they are described as bacteriostatic. The general perception both in the pharmaceutical industry and antibiotics that kill must be better than those that do not. Usually bactericidal drugs are to be preferred than bacteriostatic drugs, especially in immuno-suppressed patients, but the over-riding consideration in assessing an antibiotic drug is experience of its efficacy in clinical practice. An important factor in the cure of infection is the patient's own defence system; antibiotics cannot cure or prevent infection in the absence of adequate numbers of functional white cells in the blood. Antibiotics are well able to arrest the growth of the bacteria sufficiently for the patient's white cells to be able to eliminate the infection.

Improved activity or better penetration
The drug designer wants to improve the activity of the compound, so that a lower concentration of a potentially toxic chemical is administered. This assessment of activity is usually made in laboratory studies to measure the concentration of antibiotic required to inhibit the growth of the bacterium. The aim to reduce the physical level of drug in the body is a laudable one, however, the increase in antibacterial activity also may mean an increase in toxic side-effects. The chemical nucleus may also be altered to increase the penetration of the antibiotic, to increase the concentration at the site of infection. This might mean alteration of the nucleus to increase absorption of an oral drug from the intestine, for example the introduction of amoxycillin in place of ampicillin, primarily because of its penetration through the intestine wall. The alterations may also improve the penetration of the antibiotic into the bacterium itself. As we shall shortly see, alterations of the penicillin nucleus can radically improve the penetration into gram-negative bacteria.

One drug or two?
If a straw poll was taken by most patients and even some prescribers, would it be better to give two antibiotics rather than one, the agreement would be that more must be better. There are elegant theories why certain drugs should work in combination and why some should be antagonistic. These observations are either derived from esoteric laboratory experiments or are often just theoretical and most are simply irrelevant in the clinical situation and serve only to promote confusion amongst clinicians. For example, it is believed that is you administer a bactericidal drug with a bacteriostatic antibiotic, the latter reduces the activity of the former. This is easy to prove in laboratory experiments but is often unobserved in clinical practice. Indeed, a common treatment of meningitis, the most critical of community acquired acute bacterial infection, has been the combination of chloramphenicol and ampicillin. The theory that bacteriostatic chloramphenicol would reduce the efficacy of ampicillin paled in comparison with the reality that the two drugs greatly increased the possible spectrum of activity that the two drugs together provided so that all possible pathogens would be catered for. Combinations of drugs may also be used in order to prevent the emergence of drug-resistant strains, e.g. in treatment of tuberculosis. This theory works well with the enclosed population of organisms that mycobacteria represent and where resistance is known to emerge during prolonged treatment of an individual patient. The same arguments do not apply where resistance can emerge in a group of organisms and can be freely transmitted from one bacterium to another by movable genetic elements such as plasmids. One of the most common use of combinations is to overcome a resistance mechanism; sometimes one component of the combination might be sacrificial to the resistance mechanism, blocking it so that the other drug can work unimpeded.

There are also many disadvantages in giving combinations of antimicrobial drugs when one drug would suffice and some combinations of drugs do show antagonistic effects in clinical practice.

Combating resistance

The main impetus for altering the nucleus has been to overcome resistance. Antibiotics are conveniently classified, quite unjustifiably, into generations; in almost every case a new generation is introduced to overcome resistance to the previous generation. The basic nucleus is re-evaluated to add a functional chemical group so that the resistance mechanism which emerged to the original antibiotic cannot cope with the variant.

THE ANTIBIOTICS
Inhibitors of cell wall synthesis
β-*lactams*

In the development of antibiotics, this has the been by far the most successful and adaptable group. They all act in the final step of cell wall synthesis in which strands of peptidoglycan are cross-linked via a pentapeptide side chains the end of biosynthetic pathway. All these antibiotics contain a β-lactam ring and antibiotics resemble the terminal D-alanine-D-alanine of the pentapeptide and bind covalently to the active site of the transpeptidase enzyme so inhibiting it. This is the step required for cross-linking the polysaccharide chains in cell wall peptidoglycan. Individual β-lactam drugs also interact with a number of other proteins at the cell membrane that are termed penicillin binding proteins (PBP's). The number and types of PBP in a cell varies among species. They are selective because human cells do not have a cell wall so remain unaffected.

Penicillins

This was the first group of antibiotics and during the first 10 years of their use most drugs used were obtained by the traditional fermentation methods and improvements in technology were aimed at increasing the yield. This was often achieved by altering the fermentation conditions or the Penicillum strain used. Improvements in stability, particularly to stomach acid, were achieved by similar changes so that the penicillin nucleus had a phenoxy group (Penicillin V), which conferred acid stability, rather than the original benzyl group of the original penicillin G. This method of improvement to antibiotic design was speculative and very laborious. It soon became obvious that all penicillins were of the same basic design and it would be much more efficient to start from the basic and create new "designer" antibiotics each with particular properties that were needed. The basic nucleus of penicillin comprises a five-membered thiazolidine ring and a β-lactam ring and all penicillin derivatives are constructs from that nucleus. In the 1950s, it was

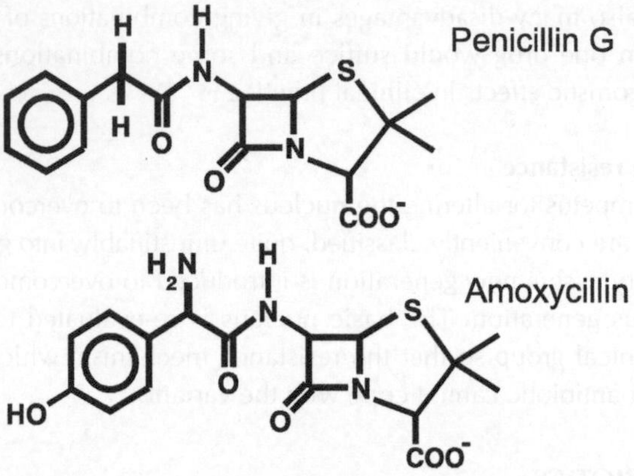

H H
Penicillin G

H₂N H
Amoxycillin

HO

COO⁻

The original Penicillin G and its semi-synthetic derivative Amoxycillin

demonstrated that Penicillin G, obtained by fermentation, could be reduced to the basic nucleus, 6-aminopenicillanic acid by treatment with an amidase, which cleaved the benzyl group. The 6-aminopenicillanic acid can then be chemically modified to change the properties of the penicillin. The first was to add a methyl group to produce methicillin. The original penicillin G had been rendered ineffective against staphylococci because they started to produce β-lactamases which destroyed the antibiotic. The addition of the methyl group meant that the penicillin could no longer bind to the β-lacta-mase and thus could not be destroyed. The addition did not increase the number of bacterial species that could be treated not did it confer the ability to overcome the stomach acid. It did not increase the activity of the peni-cillin, in fact it had the opposite effect and reduced it some 30-fold. The addition of the methyl group was merely a device to overcome the devastat-ing effects of emerging penicillin resistance which had become rampant in the 1950s. Methicillin had the disadvantage that it is acid labile so can only be administered by injection; a similar derivative flucloxacillin was acid stable so could be given orally.

The production of methicillin had been a major achievement of the Beecham pharmaceutical company. It enabled the penicillins to be used extensively against the staphylococci. Penicillins remained the predominant therapy against staphylococci for another 15 years, until the first emergence of methicillin-resistant *Staphylococcus aureus*. This heralded the end of the predominance of penicillins against this most insidious of hospital pathogens. The scientists at Beechams experimented further with chemical additions to 6-aminopenicillanic acid and found that if an α-aminobenzyl group was chemically added to the 6 position, the drug had remarkable alterations in properties. It was approximately the same efficacy as the original penicillin G, weight for weight. However, the substitution rendered the molecule stable to acid attack so that it could pass through the gastric acid relatively intact. This meant that is could be given orally and, more importantly, it could penetrate the outer membrane of gram-negative bacteria so the half of the bacterial kingdom that were previously impervious to penicillins were now vulnerable, at least in theory. Ampicillin was really remarkable when it was introduced in the 1960s, it covered infections caused by both gram-positive and gram-negative bacteria. In fact it was to be preferred for some of the gram-positive infections, such as those caused by the Enterococcus or *Listeria* genus. It has the disadvantage that it was poorly absorbed in the gastrointestinal tract; at the time this was only a minor inconvenience and ampicillin was instantly recognised as a major advance in antibacterial therapy and was used extensively throughout the world. In some parts of the world it still is, because now it is available generically, it is very cheap and within the budget of many developing countries. It has the disadvantage of instability particularly if stored badly and this may be exacerbated in tropical countries.

The poor absorption meant that significant bacteria in the large bowel were challenged with low concentrations of ampicillin, this is an optimum environment for the selection of resistance and, particularly amongst the Enterobacteriaceae, resistance developed quickly, especially as we shall see later by the TEM-1 β-lactamase. It is no exaggeration to state that this single enzyme has been responsible for the development of more antibiotics than any other. The initial response by Beecham was not to seek a drug that was resistant to the effects of the TEM-1 β-lactamase but rather to alter penicillin substituent so that the molecule would be absorbed more readily, leaving less residual antibiotic in the large bowel so weakening the environment for the selection of resistance.

It was achieved simply by adding a hydroxyl group. It improved the absorption of the penicillin to 80% or more. This not only had beneficial effects in discouraging the development of resistance but it also meant that, weight for weight, more active drug reached the site of infection. It is estimated that the concentration of active amoxycillin is between two and three-fold greater than those of ampicillin. Once the drug was absorbed, the pharmacokinetics were the same as was the spectrum of activity. These ampicillin/amoxycillin group of penicillins has become the most widely group of antibiotics around the world.

These alterations, while improving efficacy against gram-negative bacteria did not allow treatment of the one gram-negative bacteria which was considered, at the time, to be the major threat to successful bacterial management in hospitals, *Pseudomonas aeruginosa*. Addition of an a-carboxyl group at the 6-position, resulted in a molecule that could penetrate this resistant species. This is now prepared as an ester, Inanyl carbenicillin, which is inactive in its own right but should help it pass through stomach acid. Once the molecule is in the gut or is absorbed into the blood stream, esterases render the molecule active. The injectable version was a mainstay of anti-pseudomonas therapy for very many years from the 1960s onwards. Like ampicillin/amoxycillin, the substitution only broadened the spectrum of the antibiotic, it did not give it any greater capability to overcome resistance. *Pseudomonas aeruginosa* can also carry the genes for the TEM β-lactamases, indeed some believe that the may have initially disseminated in this species, and carbenicillin confers no inherent resistance to this ubiquitous resistance mechanism.

This limitation was also true for the substituted ureido-penicillins developed by Bayer. These had a very broad spectrum and were particularly useful against *Pseudomonas aeruginosa*. Azlocillin and, particularly, mezlocillin were widely used a wide range of hospital infections. Mezlocillin could not be absorbed orally so, as an injectable therapy, it remained a main-stay in the control of hospital infection until the early 1980s. It also allowed control of other infections, particularly severe lower respiratory tract infec-tions. There was some success in the treatment of *Pseudomonas aeruginosa* infection in cystic fibrosis patients, a group for whom this type of infection had been life-threatening.

All the anti-gram-negative penicillins were vulnerable to the rapidly-spreading plasmid-encoded β-lactamases. Attempts were made to maintain or even improve the spectrum. Two semi-synthetic penicillin derivatives, piperacillin and ticarcillin, had increased activity over the ureidopenicillins. They were truly broad spectrum with very good penetration into most gram-negative bacteria. These penicillin-derivatives had equivalent activity against gram-positive and gram-negative bacteria, acting as well against *Pseudomonas aeruginosa* infection as those caused by *Klebsiella* species. Piperacillin also conferred some insusceptibility to the β-lactamases of the *Neisseria gonorrhoeae*, the causative organism of gonorrhoea but the vulner-ability of the rest of the penicillins to the major β-lactamases, was still evident. In fact, there was an inherent weakness in the use of the penicillin nucleus for overcoming resistance. The only point of substitution was at posi-tion 6; this had been shown with methicillin where the substitution to over-come staphylococcal β-lactamases had reduced the activity. The paradox that had become clear was that substitution at the 6-position could be made to improve the spectrum or overcome resistance, but not both. This was a major disadvantage for treatment of gram-negative bacteria because a modification was required to broaden the spectrum to include gram-negative bacteria so one that included and overcame β-lactamase-derived resistance was a virtually impossible goal.

β-lactamase inhibitors
The future of penicillins was bleak until Beechams came up with a unique strategy, give a sacrificial drug along with the active therapy. They had a

number of β-lactamase-inhibitors that, although efficient at binding at the active site of common β-lactamases, particularly the TEM group, they were virtually without anti-bacterial activity at all. They did not bind to the penicillin binding proteins, the normal target of all the β-lactam antibiotics. The most successful of this group was clavulanic acid. This is a β-lactam compound but instead of a thiazole ring, the β-lactam ring is bonded to a five-membered oxygen ring. Clavulanic acid has an advantage over the penicillins that, weight for weight, it binds to the β-lactamase more quickly and tightly than the penicillins. This means that if equivalent levels of clavulanic acid and the penicillin are challenged by the β-lactamase, then more clavulanic acid molecules will bind to the available β-lactamase active sites, thus rendering them unavailable to the penicillin molecules. Clavulanic acid has one further advantage, although it is structural analogue it does not behave in a classical inhibitory manner. Most structural analogues are competitive inhibitors. This means that they may bind preferentially to the active site compared with the normal substrate but this binding is reversible. So if the concentration of the substrate is increased, it will start replacing the inhibitor at the active site. As the substrate is hydrolysed and the products break away from the enzyme, the active site becomes available for either another substrate molecule or inhibitor. If the former is high concentration, then more substrate will be hydrolysed. Clavulanic acid works in a slightly different manner. It is known as a suicide inhibitor; about 75% it functions as a classic competitive inhibitor where no modification of the molecule occurs but for the remainder of the time, it becomes suicidal. Its binding to the active site initiates an attempt at hydrolysis. This creates an intermediate which remains stuck in the active site. It cannot break away from the active site and thus effectively removes its availability permanently to destroy penicillin molecules. If we take the rough approximation of a 25% suicidal rate, it means that during one of the first four challenges an active site faces with clavulanic acid, it will be effectively destroyed.

This is a very important characteristic for this type of inhibitor because it provides much greater flexibility in the way the drug can be administered. The principle is that when the penicillin and the β-lactamase-inhibitor are Co-administered, the inhibitor will entangle the β-lactamase while the intact penicillin destroys the bacterial cell by binding to the penicillin binding proteins. If clavulanic acid was a straight-forward competitive inhibitor, then its efficacy would depend on the relative amount of penicillin surrounding the bacterium. If this was high, perhaps because the penicillin reached the site of infection before the clavulanic acid, then many penicillin molecules would be inactivated before clavulanic acid could "rescue" the situation. If the inhibitor has a 25% suicide rate, then far fewer molecules are needed to limit the damage by the β-lactamase. It also means that if the inhibitor is excreted more rapidly than the penicillin, it will continue to inhibit the β-lactamase even though the surrounding concentration of the inhibitor has fallen to apparently insufficient levels.

Clavulanic acid has been successfully co-administered with amoxycillin (co-amoxiclav; Augmentin®) and this has become one of the most successful antibiotic preparations, particularly popular for the treatment of respiratory infections. Clavulanic acid is also used with ticarcillin (Timentin®), which exploits the better penetration and broader spectrum of ticarcillin over amoxycillin and is used to treat severe hospital infections caused by β-lactamase-producing bacteria.

Two other β-lactamase inhibitors are available for clinical use, sulbactam which is combined with ampicillin (Unasyn®) and tazobactam combined with piperacillin (Tazocin®). Both these inhibitors contain a β-lactam ring and function in the same way as clavulanic acid, acting as suicide inhibitors of β-lactamases. Sulbactam is available in some countries on its own so that the clinician can decide with which β-lactam to administer it.

Cephalosporins

When Brotsu discovered the first cephalosporins 50 years ago, their initial advantage over the penicillins was considered to be that they had activity against gram-negative bacteria. In fact, it has turned out to be their remarkable ability to overcome antibiotic resistance mechanisms, particularly β-lactamases, that has proved to be their forte. The cephalosporins contain a β-lactam ring that is attached to a six-membered dihydrothiazine ring, a cephem nucleus. This is similar to the five-membered ring of the penicillins but it gave much greater flexibility in the modification of the molecule. Like the early penicillin, the early cephalosporins were simply fermentation products. However, these were not considered to be useful and did not enjoy wide acceptance. The breakthrough for the cephalosporins came with semi-synthetic products which were achieved in the same way as the semi-synthetic penicillins. After fermentation, the cephalosporin molecule is reduced to 7-cephalosporanic acid. The 7 position is at the same position in the cephalosporin β-lactam ring as the 6 position is in the penicillins. Chemical substitution of the 7-position can produce an enormous array of properties but it is not the only position that substitutions can occur. The 3-position on the cephem ring is also available to chemical modification. In general, but not exclusively, alterations in the 7-position change the function, mainly altering the spectrum, increasing the capability to inhibit some gram-negative species, or β-lactamase stability whereas alterations in the 3 position change the pharmaco-kinetic properties, producing changes in absorption, protein-binding, metabolism and excretion. This makes the cephalosporins the most adaptable of all the antibiotics that we currently possess. They also have another inherent property, they are larger than penicillins, this means that they are less capable of binding to the active sites of β-lactamases that have emerged during penicillin challenge. This does not mean that they cannot bind but do so less readily. The early cephalosporins, conveniently called first-generation, were quite simple modifications to increase the spectrum. These are exemplified by cephalothin and cefazolin. Cephalothin is parenteral with a broad-spectrum of activity, both against gram-positive and gram-negative bacteria. However, it does not have significant activity against methicillin-resistant *Staphylococcus aureus* so, as an injectable drug, it has very limited use within hospitals. Cefazolin has the same constraints as cephalothin against staphylococci but it is generally more active against the gram-negative bacteria. These drugs found themselves a niche in surgical prophylaxis where they are given approximately an hour before surgery. They have a short half-life and reach the peak concentrations within an hour after an administration.

Three other drugs which fall into this category of first-generation cephalosporins are cefaclor, cephalexin and cephradine. These three cephalosporins could all be used orally and possessed limited insusceptibility to the TEM group of β-lactamases. Cephradine and cephalexin are very similar drugs with good anti-gram-positive activity and can also be useful against some gram-negative infections. Their oral availability would make them good candidates for treatment of community-acquired respiratory tract infections. Indeed, they may be useful against pneumonia caused by *Streptococcus pneumoniae* but they are not effective against the major cause of bronchitis, *Haemophilus influenzae*. Cefaclor is the exception in this group in that it possesses useful activity against *Haemophilus influenzae*; in fact it is usually more effective against gram-positive infections as well. This enabled cefaclor to become a drug of choice for the treatment of lower-respiratory infections. As it is a β-lactam, it was considered safe as well as effective. The predominance of lower-respiratory infection amongst the treatable bacterial infections ensured that cefaclor become a major drug and it dominated cephalosporin usage in the United States for many years. It shared with cephalexin and cephradine some β-lactamase stability and this was also evident with β-lactamase-producing *Haemophilus influenza*. This organism did, however, develop other β-lactam resistance mechanisms which cefaclor was not effective against. Indeed cefaclor usage may have been instrumental in selecting them.

These early cephalosporins were also showing their limitations, they were not able to deal with severe hospital infections. They had limited spectrum particularly when it came to treating the causative penicillin-resistant gram-negative bacteria. A desire to increase the spectrum initially came with the usually compromises, increasing the spectrum and β-lactamase stability could not be achieved with acid stability, so new modifications could only result in parenteral drugs. The most successful of this new second generation of cephalosporins was cefuroxime, produced initially by Glaxo. This drug was particularly successful against pneumonias, including those caused by *Streptococcus* and *Haemophilus* species, including the penicillin-resistant Haemophilus influenzae. This cephalosporin was also effective against another respiratory pathogen *Moraxella catarrhalis*. Cefuroxime became a mainstay to control hospital infection but cefuroxime was ineffective against anaerobes and to increase the ability to treat these bacteria, another cephalosporin nucleus was manufactured from which were derived the cephamycins. Cefoxitin has been the most widely used of this group although there has been greater usage of cefotetan in recent years. Cefotetan is a little less active than cefoxitin but has a longer half-life.

The second generation cephalosporins always had the limitation that they were not able to treat the most severe hospital infections and these still had to be treated by the much more toxic aminoglycosides. Although generally insusceptible to the TEM β-lactamases, the second generation were still susceptible to some of the β-lactamases, particularly the chromosomally-encoded enzymes. Their wide usage in hospitals during the 1980s came at a time when the philosophy of hospital treatment changed; patients should no longer be treated for long periods of time within the hospital environment but should be released as soon as possible and continue therapy at home. This has logistic problems if parenteral drugs are to be used. Antibiotics with short half-life are impractical unless the patient can be trained to administer the drugs themselves, usually an inappropriate option. The alternatives are either to make the drug available orally or increase the half-life so that community healthcare workers have to visit, at most, once a day. Some second generation cephalosporins have been prepared in oral version. To carry them through the gastric acid, they have been esterified. Cefuroxime axetil or cefpedoxime proxetil are inert until they have passed the stomach and then, after enzymic cleavage, they are released as active drugs. So a patient who has been treated in hospital can be moved back to the community and still continue treatment. Less success was achieved with increasing the half-life.

The safety of the cephalosporins ensured their wide usage; however, there did have limited spectrum and they were still susceptible to some β-lactamases. During the late 1970s, further modifications were made which mostly improved their penetration into gram-negative bacteria. Some retained activity against gram-positive bacteria but the main threat was still considered to be gram-negative bacteria. These alterations ensured that some of these latter cephalosporins, such as ceftazidime, penetrated bacteria which had traditionally been considered resistant, most notably *Pseudomonas aeruginosa*. Some cephalosporins, such as cefotaxime, penetrated the Enterobacteriaceae much more rapidly and this rapid entry resulted in an increased capability to overcome β-lactamase resistance mechanisms. The problem was that these differing capabilities were often exclusive of

Structure of two important later-generation cephalosporins

each other and that increased capability against *Pseudomonas aeruginosa* was achieved with limited proficiency against the Enterobacteriaceae and *vice versa*. Although both types of these third-generation cephalosporins had increased stability to the plasmid β-lactamases and most of the chromosomal enzymes, it could not reasonably be supposed that a single third generation cephalosporin had a sufficiently broad spectrum to cover both Enterobacteriaceae and *Pseudomonas aeruginosa*, though some were perceived and were marketed with these claims. This is inappropriate perception and use of third-generation cephalosporins is probably responsible for the rapid emergence of extended-spectrum β-lactamases.

As their use increased, the half-life of early third-generation cephalosporins was considered to be a significant limitation and further variants were developed with considerable increases in half-life, to as long at 33 hours which ensured that the drug need be administered only once a day. The best example is ceftriaxone, which in terms of activity is very similar to cefotaxime; however, its extended half-life ensured its popularity so that it became the most widely used drug in its class. The other limitation on this extended-spectrum cephalosporins was that there were invariably injectable. This was not an undue problem in much of Europe but it did limit the acceptance of these drugs in the United Kingdom and the United States as this effectively meant that they were restricted to use in hospitals. Oral versions were introduced in the 1990s, the first was cefixime which had extended activity against most of the Enterobacteriaceae but not much against *Pseudomonas aeruginosa* and except for the latter, it resembled ceftazidime in profile against these bacteria. This may be because it did not have particularly fast penetration through the outer membrane. It was also ineffective against most staphylococci and *Enterobacter* spp. Cefixime has a relatively long half-life but was poorly absorbed from the gut, only about 50% and this resulted in gastrointestinal upsets during therapy. Cefdinir, another oral third generation cephalosporin, had an activity more akin to cefotaxime and penetrated the cells more quickly than cefixime. Cefdinir had some potential against Enterobacericeae that had become difficult to treat with other β-lactams.

It has become clear that the faster that a cephalosporin could pass through the outer membrane and reach the penicillin binding proteins, the more effective it was likely to be. Thus improved penetration became the essential property of new cephalosporins. Two cephalosporins were developed as zwitterions, that is they carried both negative and positive charges. This gave them much quicker entry into the cell and could bring bacteria, which possessed β-lactamases and had resisted previous cephalosporins, under their control. These fourth generation cephalosporins, cefepime and cefpirome, have just been released for clinical use and we shall have to see whether rapid access to the target is a significant advantage.

Monobactams
In the early 1980s, remembering Ehrlich's dream of a magic bullet, the desire was to seek an antibiotic that specifically inhibited the pathogen and leave all other bacteria unaffected. This goal was nearly reached with the monobactams. These are β-lactam rings without an attached side-ring. There is only one clinical example, aztreonam and it is only active against gram-negative species and shows no activity against gram-positive bacteria. Aztreonam has a side chain, at the equivalent of the 7-position of a cephalosporin, similar to ceftazidime with a concomitant similarity of properties Unfortunately its use relied on the assumption that an infection was caused by a single bacterial species and that there were neither mixed infections nor was there a possibility of super-infection. It was hoped that this drug would be useful in areas where gram-negative infections were particularly abundant, in the treatment of gonorrhoea and *Haemophilus influenzae* infections of the chest. In the treatment of sexually transmitted diseases it was shown to have some role, but the chest is more complicated and there were problems with its use if there were gram-positive streptococci in the chest.

Carbapenems
There are natural β-lactams that rapidly penetrate gram-negative bacteria, these are the thienamycins. Synthetic variants of these natural β-lactam com-

pounds has resulted in the carbapenems. The nucleus of the carbapenems is similar to that of penicillins, with a five-membered side ring but differs in the replacement of sulphur by carbon. This is a difficult group of compounds to synthesise and they are inherently unstable. This makes them extremely expensive; however, like the fourth-generation cephalosporins, these are zwitterions and they have extremely rapid penetration into gram-negative bacteria. So rapid and easy is their entry through the outer membrane that very few gram-negative bacteria are inherently resistant and the carbapenems have the broadest spectrum of activity of any of the β-lactam family. They are active against gram-positive as well as gram-negative bacteria, showing comparative efficacies against aerobes and anaerobes. The first of these injectable β-lactams was imipenem and this drug was soon considered as the final defence against almost all serious, hospital-acquired gram-negative infection except *Stenotrophomonas maltophilia* and *Burholderia cepacia*. Imipenem has a significant disadvantage, in that it can be metabolised by peptidases so it always administered in conjunction with a renal dihydropetidase-I inhibitor, cilastatin. It is considered not safe enough to give to give to children. The only other carbapenem currently available is meropenem which has the not inconsiderable advantage that it has decreased nephrotoxicity and is stable in the presence of renal dihydropetidase-I. It has, therefore, been used in children particularly for the treatment of meningitis as it was thought be associated with a low risk of fits. Despite the fact that it was released many years after imipenem, meropenem was readily accepted in the United Kingdom and is the major carbapenem used in hospitals in this country.

The model of substituting the sulphur of the β-lactam side-ring with a carbon proved less successful with cephalosporin nucleus. These are known as carbacephems and the most widely used example is loracabef. Loracarbef is structurally very similar to cefaclor and retains the capability to be given orally. Unfortunately the substitution with carbon in a six-membered side-ring is not nearly as beneficial as it was the five-membered ring. The drug does not penetrate as fast and does not have anything like the same spectrum of activity. It does not inhibit *Pseudomonas aeruginosa* or *Acinetobacter baumannii*.

Glycopeptides

Vancomycin

Vancomycin is a glycopeptide antibiotic which is obtained from the natural fermentation of Streptomyces orentalis. In the 50 years, since its first discovery, this drug has remained largely used in an unmodified state. It has a narrow spectrum of activity against gram-positive bacteria acts by binding peptides containing D-alanyl-D-alanine during bacterial cell wall synthesis.

Vancomycin prevents D-alanyl-D-alanine interacting with the peptidoglycan synthetase, so preventing the polymerisation of UDP-N-acetyl-muramyl pentapeptide and N-acetylglucosamine into peptidoglycan. Its unique activity against gram-positive bacteria was ignored for many years in competition with more active and safer antibiotics, especially methicillin. However, as *Staphylococcus aureus* became more resistant, particularly to methicillin and gentamicin, the options for treatment became very limited and vancomycin enjoyed a revival; so much that it now, as one of the very limited range of drugs available to treat resistant *Staphylococcus aureus*, is a very successful product for its manufacturers.

Vancomycin was used extensively against *Enterococcus* species. As this genus rose to prominence in the Intensive Care Units, mainly because of it predisposition to resistance, the number of antibiotic options reduced rapidly and vancomycin was temporarily very effective. Vancomycin-resistance has all but halted its use against enterococci. It was also used against the gram-positive anaerobic bacterium, *Clostridium difficile*. This bacterium is often associated with antibiotic-associated colitis but responds well to vancomycin. Vancomycin was used to control the unwanted destructive effects of the use of other antibiotics.

The main difficulty with vancomycin has been toxicity. Early preparations contained significant concentration of contaminating chemicals and these were certainly a contributory factor to the side-effects. Better purification procedures in manufacture alleviated this problem but did not remove it. Careful attention to dose, particularly in patients with impaired renal function, has significantly reduced the number of reported side-effects. This often means monitoring the amount of the drug in the blood to ensure that it is being excreted sufficiently quickly. However, this glycopeptide has always been associated with toxic effects and safer alternatives were sought.

Teicplaninin is a glycopeptide obtained from the fermentation of *Acintoplanus teichomyceticus*. In reality, teicoplanin is a mixture of related glycopepetides each with similar heptapepetide bases and an aglycone containing aromatic amino acids and D-mannose and N-acetyl-D-glycosamine. Unlike vancomyicn, it has a fatty acid as an acyl substituent which makes it much more lipophilic. Teicoplanin has a similar action to vancomycin on the bacterial cell. It is considered to be slightly more active than vancomycin and it has an improved pharmacological profile.

The main advantage of teicoplanin is its improved safety profile. It has much the same spectrum of activity as vancomycin but does not usually require the same monitoring for toxic side-effects or blood levels for drug accumulation. The glycopeptides have remained prominent because of the emergence of resistance to almost all other competing drugs. There are some minor differences in resistance profile but these differences are difficult to exploit.

Bacitracin

Bacitracin is an antibiotic that ranks by itself. It comes from the fermentation of *Bacillus lichenoformis (Bacillus subtilis)*. Like penicillin it contains a thiazolidine ring attached to an amino acid. It contains no β-lactam ring. It acts at the stage of linear polymerization of peptidoglycan synthesis, particularly by inhibiting conversion of phospholipid pyrophosphate to phospholipid. This is essential for the regeneration of the lipid carrier required. It has not been extensively used because it is usually considered too toxic for anything other than topical use.

Polymyxins

There are a series of related cyclic basic polypepetides (Polymyxin A - E) which are produced by *Bacillus polymyxa*, only Polymyxin B and, more latterly, Polymyxin E are used for clinical treatment as the other three are considered to be too toxic. They are not classic cell-wall inhibitors as are the previous compounds but rather they act on the outer membrane. Parts of the molecule have an affinity for water (hydrophilic) whereas the rest actively repels it (hydrophobic). This hydrophilic property allows the molecule to bind to the anionic phospholipids then the hydrophobic regions enter the hydrophobic phospholipid region of the outer membrane. The polypeptide rings bind to the anionic phosphate groups of the membrane. The molecule works as a detergent, this surfactant action disrupts the membrane causing increased permeability both in and out of the cell, the latter resulting in the loss of proteins and nucleic acids. There are similar membranes in mammalian cells and the binding of Polymyxin to these reduces its selective toxicity and it is known to have significant nephrotoxicity. Its ability to effectively increase the permeability of the cell has meant that it could be used in combination with other antibiotics, which would not normally have penetrated the outer membrane. Trimethoprim was administered with Polymyxin B for the treatment of *Pseudomonas aeruginosa* infections. *Pseudomonas aeruginosa* would not normally have been treatable with trimethoprim because the drug does not penetrate; however, with the increased permeability associated with Polymyxin B, the trimethoprim can pass through to the cell cytoplasm and bind to its normal target.

It is known that there are other peptides that have similar antibacterial effects. They are in various stages of development and, as yet, none have reached clinical use. Although there may be some toxicity problems with them, there are many proteases (enzymes that destroy proteins) in the body and some of these compounds may have difficulty in reaching their target bacteria intact.

Inhibitors of protein synthesis

In the search for antibiotics as natural products, inhibitors of protein synthesis have been by far the most prolific. The problem with inhibitors of this essen-

tial process is that mammalian cells have a very similar mechanism for synthesising proteins, so most of the natural inhibitors of protein synthesis are
equally effective against mammalian and bacterial systems.

Aminoglycosides

The most successful group of antibiotics in this group have been the aminoglycosides. It is questionable whether they should really be included because
the exact mode of action is not currently well understood, despite the fact
that these drugs have been available for over 50 years. It is known that
aminoglycosides interact with the bacterial 30S but not mammalian 80S ribosome subunit and are thus selective in their activity. Aminoglycoside binding
to the bacterial ribosomes can have a number of effects, which include premature termination of the protein chain formation and misreading of the
genetic code so that the incorrect proteins are formed. It has been presumed
that the resulting inadequate production of vital proteins has disruptive
effects on many essential bacterial functions leading to cell death; however,
cell death is not usually caused by the inhibition of protein synthesis and it
has been suggested that the aminoglycosides may have other targets, particularly in the synthesis of DNA, inhibition of which is always associated with
rapid cell death. The problem is that almost all the studies on the action of
the aminoglycosides were done many years ago and only with the first
member of the group streptomycin and what was applicable to streptomycin
may well not be relevant to other aminoglycosides.

We have already seen that the first aminoglycoside came from a fermentation product of *Streptomyces griseus*, but this was a fairly toxic product
and soon fell into disuse when alternative antibiotics were available for the
treatment of common infections, though it remained prominent for anti-
tuberculosis therapy. The group became popular again with the development
of kanamycin which was used on hospital gram-negative infections.
Unfortunately it had little activity against *Pseudomonas aeruginosa* and was
soon superseded with gentamicin. This drug was found to be exceptionally
useful in the treatment of severe hospital-acquired infection, particularly
those caused by gram-negative bacteria. It could be used in combination
with penicillins and vancomycin for the treatment of gram-positive infections
and was widely used against *Staphylococcus aureus* before multi-resistant
strains emerged. It also suffered from problems of toxicity but the infections
that it was used to treat were usually life-threatening. Gentamicin was still
not as effective against the *Pseudomonas aeruginosa* as it was against other
gram-negative species and the subsequent introduction of tobramycin overcame this. Tobramycin is more active against *Pseudomonas aeruginosa* even
when the strains are sensitive to gentamicin. There were problems also with
resistance and to overcome these semi-synthetic derivatives were con-
structed, the most successful was amikacin which is a semi-synthetic deriva-

tive of kanamycin and can overcome most aminoglycoside modifying enzymes capable of conferring aminoglycoside resistance. Unlike kanamycin, it can also be used to treat *Pseudomonas aeruginosa*. Netilmicin is a semi-synthetic derivative of sisomycin. This modification is to prevent its inactivation by bacterial enzymes responsible for resistance. So netilmicin is used primarily against infections known to be gentamicin-resistant. Interestingly, as we shall later, resistance has not been an overwhelming problem for the aminoglycosides and there has not been major development of semi-synthetic derivatives as there has been for the β-lactams.

The aminoglycosides are poorly absorbed from the gastrointestinal tract and are therefore not given orally; the only exception is neomycin which is given orally for gastrointestinal infections and exploits the poor absorption from the gut. The nephrotoxicity of this class of drugs has ensured that neomycin usage has been very limited.

Macrolides
This class of drugs is only other group of protein synthesis that still enjoy wide acceptance. Erythromycin was the first of this group to be used clinically and is a natural product derived from the fermentation of *Streptomyces erythreus*. It is an extremely large with a 14-carbon ring structure and it has a narrow spectrum of activity and, when it was introduced, in 1952, it was used primarily against streptococcal infections. Erythromycin is an inhibitor of protein synthesis, binding to a single site on the 50S bacterial ribosomal subunit. The mechanism of erythromycin action is thought to stimulate the dissociation of the peptidyl-transfer RNA from the ribosomes. This has the net result of preventing translocation of the peptidyl-transfer RNA after peptide bond formation. Erythromycin does not bind to mammalian 80S ribosomes and thus it is selective. Indeed it is consider to be a relatively safe antibiotic and has been used extensively in the United States. Its ability to kill bacteria varies with different species, it does not kill *Staphylococcus aureus* but does cause death amongst the streptococci.

Erythromycin has only limited activity against a few gram-negative species including the respiratory pathogens *Haemophilus influenzae*, *Moraxella catarrhalis*, *Legionella pneumophila* (causative agent of Legionnaires' disease), *Bordatella pertussis* (causative agent of whooping cough). It is also useful in the treatment of gonorrhoea but generally it is less effective against the gram-negative bacteria and, in an environment such as the lung, where both gram-positive and gram-negative bacteria can proliferate together, a broader spectrum would be favourable. Despite the discoveries of other naturally occurring macrolides, erythromycin remained the member of this class in clinical use for 40 years. Developments of the macrolides came in the late 1980s with semi-synthetic modifications to improve the spectrum, increase the half life, improve the pharmacokinetics and provide better gastrointestinal tolerance. Indeed, now only erythromycin is used as injectable drug, all the remaining macrolides are oral preparations.

Azithromycin is a 15-membered ring, called an azalide, with a methyl-substituted nitrogen at position 9. This macrolide-like compound has a very similar mode of action to erythromycin but the modification improves its action against gram-negative bacteria, especially *Haemophilus influenzae*, thus this drug has proved very useful for the treatment of common respiratory infections.

Clarithromycin is a semi-synthetic derivative of the 14-ring nucleus so is a true macrolide. Clarithromycin could be used for the treatment of chest infections caused by *Chlamydia pneumoniae*, a difficult pathogen to treat as it is not strictly a bacterium and is thus it is not easy to find true selective therapy against it. Clarithromycin has proved particularly useful in the treatment of some of the recently discovered pathogens, *Legionella pneumoniae*, *Campylobacter jejuni* (causative organism of severe gastrointestinal infection) and *Helicobacter pylori* (causative organism of stomach and duodenal ulcers). The other advantages of the semi-synthetic macrolides are their increased serum levels and their half-lives (which can range up to 68 hours) resulting in much longer dosing intervals.

Streptogramins

The streptogramins are a class of proteins synthesis inhibitors similar to the macrolides. There are naturally occurring streptogramins but it is the combination of two streptogramins quinupristin and dalfopristin which produced a synergistic effect that killed susceptible bacteria. This combination known as Synercid® is the first injectable streptogramin, and has just been released for clinical use in the United Kingdom. This combination drug works effectively against methicillin-resistant *Staphylococcus aureus* and vancomycin-resistant enterococci. In many parts of the world, the increasing difficulty experienced to treat these two hospital-acquired infections suggests that Synercid® may become widely used.

Its exact mode of action is unknown but it is believed to act on proteins synthesis by reducing the exit channel for the growing protein resulting in accumulation of these partially formed peptides attached to transfer RNA, thus limiting the available transfer RNA for the formation of new proteins.

Lincomycins

Lincomycin is a natural product derived from the fermentation of *Streptomyces lincolnesis*. It was found in 1962 and, by 1966, a semi-synthetic modification was made by adding 7-deoxy, 7-chloro-derivative to produce clindamycin. These lincomycins bind to the bacterial 50S ribosomal subunit. They appear to bind at the same site as chloramphenicol and the macrolides but the effect of the lincomycins is to prevent initiation of peptide chain formation by binding to ribsomal subunits that are not being currently being used in proteins synthesis. They are predominantly bacteriostatic drugs although under certain conditions can be bactericidal. They are active primarily against gram-positive bacteria and anaerobes. Rare but serious side effects with these antibiotics means that they should only be used in situations where the patients can be monitored closely. Clindamycin has largely replaced lincomycin in clinical use because it is more active and has improved absorption from the gut.

The remaining groups of protein synthesis inhibitors comprise just one or two closely related drugs. They are mainly antibiotics that were found in the first wave of drugs and have not lent themselves to systematic modification. They often suffer from the toxicity problems of natural fermentation products and usually have short half-lives.

Tetracyclines

We have seen in the previous chapter that the tetracyclines were obtained originally from the fermentation of soil bacterium *Streptomyces aureofaciens* and that the active component was chlorotetracycline. This is no longer available except for topical use and active tetracycline was obtained by the catalytic removal of the chlorine atom. Thus the original basic molecule, although a modification of the natural product, was not semi-synthetic. Two semi-synthetic derivatives, doxycycline and minocycline were synthesised and were obtained by different substitutions on the four ring system that comprises the basic molecule. These semi-synthetic substitutions improved the properties rather than overcome problems with resistance. Doxycycline increased the half-life to about 18 hours from about 8 hours while minocycline has increased activity and a broader spectrum with a half-life of about 16 hours. It can even be used against some strains of *Stenotrophomonas maltophilia.*

Tetracyclines inhibit protein synthesis by binding to the 30S ribosomal subunit. They will also bind to the equivalent 40S subunit in mammalian cells but tetracyclines are such large molecules that they have to be actively transported into bacteria though a small amount may enter by passive diffusion; mammalian cells do not have an equivalent transport system so tetracyclines are selective because they do not penetrate mammalian cells. The binding to the 30S unit prevents the attachment of aminoacyl-tRNA so obstructing the addition of new amino acids to the peptide chain. Although tetracyclines have a broad spectrum of activity encompassing both gram-negative and gram-positive bacteria, aerobes and anaerobes; there are number of toxic side-effects associated with them but, for serious infection, their main disadvantage has been that their action is bacteriostatic. In general, their use has declined as newer agents have been developed.

Chloramphenicol

A protein synthesis inhibitor with which there has been even less development has been the natural product chloramphenicol. Chloramphenicol binds to the 50S ribosomal subunit of bacterial cells but does not bind to the equivalent 40S unit in mammalian cells. This interaction prevents the attachment of the amino acid-containing end of the aminoacyl-tRNA to its binding region thus preventing these compounds reacting with peptidyl transferase preventing the formation of the peptide bond. The binding of chloramphenicol to the 50S subunit is reversible so the drug does not kill the bacterium, merely stops it growing. Even so it has a broad spectrum of activity but it is considered too toxic for common infections and is only used to treat seriously ill patients in well-defined situations. It was the drug of choice for *S. typhi* infections for many years but increasing resistance is being reported.

Mupirocin

Mupirocin is a semi-synthetic derivative of a natural product obtained from the fermentation of Pseudomonas fluorescens. It is a crotonic acid ester of 9-hydroxynonanoic acid and acts as a protein synthesis inhibitor by binding to bacterial isoleucine-transfer-RNA synthetase thus preventing the incorporation of isoleucine into protein. The binding is reversible so the antibiotic is bacteriostatic. Unfortunately it is too toxic to use systemically and can only be used on the surface of the body. It is extremely active against methicillin-resistant *Staphylococcus aureus*, coagulase negative staphylococci and some streptococcal strains. It is used as an ointment to keep surface staphylococcal infections under control and to limit the carriage of *Staphylococcus aureus* in the nose.

Inhibitors of RNA synthesis

Rifampicin

The cell wall and protein synthesis inhibitors have largely derived from natural products, obtained by fermentation, and some have been modified synthetically. The only other group obtained from the same source has been the RNA synthesis inhibitor rifampicin. It is obtained by the fermentation of *Streptomyces mediterranei* and is a napthaline macrolide. It binds to the s subunit of bacterial RNA polymerase, the enzyme required to transcribe messenger RNA from the bacterial DNA. It is active against a number of different organism both gram-positive and gram-negative but it has primarily been targeted against *Mycobacterium tuberculosis*. Against this organism, the action of rifampicin is considered to be bactericidal and, as such, is thought to reduce the duration of therapy for tuberculosis. It is not bactericidal against other bacteria particularly the meningococcus and some gram-negative species. There have been no major modifications to this molecule to improve its spectrum or to overcome resistance; however, if the same producer, *Streptomyces mediterranei*, is fermented in the absence of diethylbarbituric acid, then the closely related rifamycins are synthesised. The most important of these is rifamycin SV, known as Rifocin. Although a natural product of Streptomyces mediterraneii, it is also produced by *Micromonospora chalcea* under similar cultural conditions. This antibiotic has high activity against gram-positive bacteria but only limited anti-gram-negative activity and so has proved effective against streptococcal, staphylococcal and meningococcal infections. The main disadvantage of this group of drugs is the speed at which resistance emerges and, for this reason, little development of them has been made.

The remaining antibacterial drugs are not true antibiotics as they are completely synthetic compounds. What is surprising is that there are so few of them suggesting that we are not actually very efficient at designing new antibacterial molecules. They fall into two main classes of inhibitors, those that inhibit folic acid synthesis and DNA synthesis.

Inhibitors of folic acid synthesis

Sulphonamides
We have already seen the events that led to the discovery of the sulphanil-iamide, the first of the sulphonamides. Like the cephalosporins, there has been a plethora of sulphonamides but they only affect absorption rather than conferring the ability to overcome resistance. Thus when true antibiotics, which are usually more active and safer, became established, the sulphonamides fell into disuse. The original sulphonamide, sulphanilamide is a close structural analogue of para-amino-benzoic acid. The drug binds to the active site of dihydropteroate synthetase usually occupied by para-amino-benzoic acid. This structural analogue prevents the condensation of pterin to form pteroate in the manufacture of tetrahydrofolic acid. In most cases this binding is reversible and sulphonamides are competitive inhibitors and this makes their action largely bacteriostatic. There are reports that isulphonamides can kill under certain physiological conditions but there is very little evidence to show that this occurs in the clinical situation. There is no equivalent biochemical step in mammalian cells as folic acid derivatives are absorbed intact in mammals but this does not mean that sulphonamides are especially selective. The sulphonamides have been used against a wide variety of both gram-positive and gram-negative infections. After the advent of penicillin, their use declined against gram-positive infections but they remained useful for the treatment of gram-negative infections.

There have been many attempts chemically to improve the structure and most of these variations have been produced from changes in the sulphonyl component. Alterations at this position can increase the activity and, to some extent, the spectrum. They have altered absorption, solubility, and improved gastrointestinal tolerance. These changes produced sulphonamides that were short or long acting, those that could not be absorbed from the gastrointestinal tract and those that could only be applied to the skin. None of these substitutions were effective in overcoming resistance mechanisms. On their own, sulphonamides are rarely used on their own and there is probably only one drug that we should consider, sulfamethoxazole. Sulphamethoxazole is one of the sulphonamides substituted at the sulphonyl position. It was relatively unimportant until it was marketed with trimethoprim.

Trimethoprim

Trimethoprim is really the only designer antibacterial drug. It is the sole example where a completely new chemical nucleus has been conceived and then proved widely successful in the control of infections. It was conceived by George Hitchings who was working at Burroughs-Wellcome in the United States. He knew that bacterial dihydrofolate reductase, which catalysed the final stage of tetrahydrofolate synthesis, was different in structure from the mammalian enzyme. He synthesised a number of compounds to try and exploit this difference. He found the 2,4-diaminopydimidines and started to make chemical substitutions on them. One of his early successes was pyrimethamine, which has proved successful, combined with sulphonamide sulfadoxine, for both prophylaxis and treatment of *Plasmodium falciparum* and *Plasmodium vivax*, the two main causes of malaria. Trimethoprim was found to have preference for bacterial enzymes while leaving the mammalian enzyme virtually unaffected. It is interesting to note that while Hitchings was looking for these dihydrofolate reductase inhibitors, he discovered a thymidine kinase inhibitor. This compound was considered too toxic for antibacterial use and resistance developed to it far too quickly in bacteria. The inhibitor was later called zidovudine or AZT and became the premier treatment to control the course of HIV infection. Hitchings rightly was awarded the Nobel Prize in 1990 for his discoveries of both trimethoprim and zidovudine and this was the last prize given for the discovery of a new antibiotic.

Trimethoprim was effective against both gram-positive and gram-negative infections and was very selective. It was considered to be a bacteriostatic antibiotic, after all it was only a competitive inhibitor of bacterial dihydrofolate reductase and the binding was reversible. Unfortunately all the experiments had been performed in simple laboratory media and did not emulate the action of the drug at the sites of infection. Johnty Smith and I in London and Rudolf Then and Peter Angehrn at Hoffmann-La Roche in Basle started to examine the action of the drug under conditions as close to a clinical environment as we could create. We found that trimethoprim is a drug that can kill most bacteria very rapidly and challenged bacteria exhibit all the classical characteristics of dying bacteria. This information was not known when trimethoprim was ready for marketing at the end of the 1960s. The commercial decision was to market it with the sulphonamide, sulfamethoxazole.

Personally I think that this was a poor decision and the subsequent successful marketing of trimethoprim alone bears this out. There were three reasons given for this decision. The first was that sulfamethoxazole and trimethoprim inhibit two biochemical steps in the synthesis of bacterial tetrahydrofolate reductase and that inhibition of both steps would give synergistic activity; in other words the two drugs would work more effectively together than would be expected from their individual activities. This is easy to demonstrate in laboratory tests but has never been proven in clinical efficacy studies except for the treatment of chest infections caused by *Pneumocystis carinnii* and perhaps infections caused by some strains of *Neisseria gonorrhoeae*. The second reason was trimethoprim had been considered to be bacteriostatic but, in combination with sulphamethoxazole, it killed bacteria. As we have seen, this claim was made in ignorance of the true action of trimethoprim. The third claim was that resistance to trimethoprim would develop much more slowly in the presence of sulphamethoxazole than in its absence. We shall see in a later chapter how this prediction was misguided. There are allegations that there were corporate reasons for marketing the combination that had little to do with clinical efficacy; however, the combination of trimethoprim and sulphamethoxazole was launched in the United Kingdom in 1968 with the generic name co-trimoxazole. It was an instant success and was extensively used for the treatment of common infections, particularly those of the urinary and respiratory tract. Bacteria were rapidly acquiring resistance to ampicillin, the main antibiotic used at the time to control gram-negative infections. So widespread was the use of co-trimoxazole that I have heard it suggested that every general practitioner in the United Kingdom prescribed the combination twice in every session. This is staggering successful and probably has never been matched since.

Resistance has limited the efficacy of trimethoprim to only some degree; however, it soon became clear that it would be impossible to modify the compound to overcome the emerging resistance mechanisms. After the original patents ran out at the end of the 1970s, trimethoprim was marketed on its own. Its limitation was its relatively short half-life and a modification could be made to improve this. Brodimoprim is a trimethoprim analogue with a considerably lengthened half-life but it has not been widely accepted around the world except in South America.

Inhibitors of DNA synthesis

Quinolones
The last novel chemical nucleus to be developed for the control of bacterial infection was found as long ago as 1961, with nalidixic acid. This was the first of a group of synthetic drugs known as the quinolones. Most of the drugs in this class are actually 4-quinolones because they have had an oxygen added at the 4-position to improve antibacterial activity. The origins of quinolones were actually natural products and the structure was based on quinine, a natural product from the chichona tree in South America. Quinine had long been known to effective against malaria; after all it was why British colonialists use to drink gin and tonic, the tonic water contains quinine.

Nalidixic acid was not an effective antibacterial drug. It was only active against gram-negative bacteria and could be used to treat urinary infections because only at this site did the drug accumulate in sufficient concentrations. It was rather disparagingly referred to as a urinary antiseptic. There were a series of other compounds developed based on nalidixic acid, including oxolinic acid, pipermidic acid and cinoxacin. Only the last remains in general clinical use.

The 4-quinolones can be divided into four major groups, napthridines, pydiripyrimidines, cinnolines and the true quinolones. The napthridines include nalidixic acid but the breakthrough for this class of compounds was the discovery that if a fluorine atom was added to the 6-position the activity was enhanced by up to 1000-fold and the spectrum was considerably broadened to include gram-positive bacteria. This was a remarkable breakthrough and led to the major antibacterial developments of the 1980s. This enhanced the activity of some of the sub-groups. Enoxacin is a fluorinated naphthyridine and the other highly active fluoroquinolones are enhancements of the quinolone sub-group.

All members of the class inhibit bacterial toposiomerase II or DNA gyrase, the enzyme that controls the supercoiling or folding of the bacterial chromosome DNA within the cell. They also inhibit bacterial topoisomerase IV responsible for decatentation, essentially a maintenance function in the successful packaging of the chromsomal DNA. Each fluoroquinolone has slightly different relative effects against these two enzymes and so has slightly different effects on the cell. However, the whole class is capable of rapidly killing bacteria.

The most successful member of this class is ciprofloxacin produced by Bayer AG. This drug was a revolutionary innovation and one of the most successfully marketed drugs of all time. It was highly active and had relatively few side-effects. It contained a piperazine ring which increased activity against staphylococci and *Pseudomonas aeruginosa*. This drug has become a major antibacterial in both the control of community and hospital infections.

Structure of the fluoroquinolone – ciprofloxacin

The other currently available fluoroquinolones have been less success-ful both because of reduced activity and poorer marketing. Ofloxacin has a very similar profile to ciprofloxacin and some enhanced in vitro activity against gram-positive bacteria but with reduced in vitro activity against gram-negative bacteria. It is actually a racemic mixture of D-ofloxacin and L-ofloxacin, the former is ineffective so levofloxacin, which is just the L-isomer of L-ofloxacin, has been developed in an attempt to match the activity of ciprofloxacin.

The other fluoroquinolones in current use are less effective and include enoxacin, perfloxacin and norfloxacin. They have no enhanced gram-posi-tive activity and lesser gram-negative activity. The disadvantage of the fluoro-quinolones is that their activity against gram-positive is sometimes insufficient to provide sufficient clinical efficacy. This has recently been addressed further by the development of new variants of this class specifically engineered with improved activity against gram-positive bacteria. The first of this group, which has just been released, is grepafloxacin from Glaxo-Wellcome. There are three further drugs that appear to have enhanced activity over even grepafloxacin. These are trovafloxacin from Pfizer, moxifloxacin from Bayer and gemifloxacin from SmithKline Beecham. It is likely that we shall see these drugs launched in the first two years of the millennium.

Metronidazole

This a synthetic antibacterial drug that was used as an antifungal for many years before it was realised that it had significant activity against anaerobic bacteria. The nitro group of this compound is a preferential electron acceptor that has a redox potential lower than ferrodoxin and flavodoxin, the low redox potential electron transfer proteins normally found in anaerobic and microaerophilic bacteria. The drug acts an electron sink when the 5-nitro group is reduced by the enzyme nitroreductase only found in anaerobic bacteria. This releases free radicals which require the absence of oxygen to survive. The free radicals are presumed to cause a series of single-strand breaks in chromosomal DNA, inhibiting replication, transcription and repair. The drug can kill susceptible bacteria quickly and, because it is so specific for anaerobic bacteria, it can be targeted just at this class of bacteria. There are few other good anti-anaerobe drugs and fortunately metronidazole resistance is not a significant problem.

Nitrofurans

This class of synthetic compounds are based on the 5-nitro-2-furaldehyde nucleus but have few supporters nowadays. The main member of the group, nitrofurantoin, could control a wide range pathogens in the concentrations that it reached in the urinary tract so it used to be considered important in the control of urinary tract infection. Nitrofurans can inhibit a variety of bacterial functions; they have detrimental effects on mammalian cells and their selectivity originates from their rapid excretion by the kidneys. Like metronidazole, the nitrofurans have to be activated though this does not require anaerobic conditions. They also produce free radicals and their bactericidal activity is caused by their disruption of the chromosome.

In the past 50 years there have been a massive and impressive array of new antimicrobial drugs. Actually they are based on a distressingly small number of chemical nuclei. Most of these drugs would never have been required if clinical bacteria had not been able to counteract the challenge of antibiotics, by becoming resistant.

4

Paradise lost – the emergence of resistance

In the 1949 Carol Reed film *The Third Man,* an American, with the unlikely name of Holly, arrives in Vienna to meet his friend Harry Lime; however, Lime has apparently died after a road accident. During the course of the film, Holly discovers that his friend was a profiteer from the post-war deprivations of the occupied city. He supplied the children's hospital with penicillin that was so diluted that it was ineffectual. Holly is shown pictures of the horror of the effects of this diluted drug on innocent children. The tragedy is that the antibiotic has lost its magical curative powers and Lime is branded a murderer. He gets his rightful deserts; he is justly shot in the sewers under Vienna like the rat he is portrayed to be.

Harry Lime's crime was not just that he diluted penicillin but also that he left some active antibiotic in the supplies that he sold to the hospital. Antibiotics are unique amongst all pharmaceuticals in that they do not work forever. Bacteria somehow learn to adapt to survive in their presence. This "learning" process requires mutation in the DNA of the bacteria and this takes not just time but also replication of the DNA. If a hefty, clinical dose of antibiotic is given, the speed of the antibiotic's attack is so fast that insufficient time is available for mutations or DNA replication. If the antibiotic is diluted or deficient, the drug trickles into the bacterium in insufficient concentrations to prevent either the cell dividing or DNA replication; a failure to inhibit the latter means that mutations can still occur and resistance emerges. This is the breeding ground of resistance because much antibiotic resistance comes from the use of too little antibiotic rather than too much.

MUTATIONAL RESISTANCE

To understand how resistance first emerges, the ethos of dosing levels must be examined. Bacterial infections are composed of, literally, millions of individual cells. They may be floating around in some liquid environment such as the urine or blood or, perhaps more likely and depending on the organism, they might attach themselves to a specific site within the body where they will establish the infection. Attached bacteria might form a colony, like those seen on the agar surfaces of Petri dishes. This has the advantage that it keeps the bacteria in one site, probably at the most advantageous spot to obtain the nutrients that they require. The symptoms and pathological damage are really a by-product of the bacterium's drive to reach the site of maximum nutrient or survival reward. The fact that there are so many bacteria at one site means that we should not really treat them as individual cells; this colony is more like a family of bacteria capable of helping and supporting the survival of the clan rather than that of the individual. Within the one billion bacteria that might make up such a colony there is some heterogeneity; most of them are sensitive to the antibiotic but a few are super-sensitive and a few are less sensitive. When deciding the antibiotic dose, care must be taken to ensure the drug reaches sufficient concentration to inhibit all the bacteria of the colony, including the less sensitive. If this dose is insufficient, some of the bacteria will be able to divide and mutations can occur that will give higher levels of resistance.

What are these mutations? When most bacteria are cultured in the laboratory and one billion cells are placed on the agar surface of a Petri dish containing an inhibitory dose of antibiotic, after incubation a few bacteria colonies will appear growing on the surface. This might be typically 100 colonies from that initial population of one billion. The 100 colonies derive from cells that have already undergone a mutation, which now allows them to overcome the effect of the drug. The mutation event is usually independent of the presence of the antibiotic and stems from random mutation events that occur in all genes during DNA replication. Most mutations produce lethal consequences and the bacterial cell dies, but the bacteria are part of a large "family" so this wastage is unimportant. However, the mutation may produce a small change in a gene that does not result in lethality but rather gives the bacteria an advantage under certain environmental conditions. The most immediate of these would be if the bacteria were caught up surrounded by antibiotic. The antibiotic milieu selects those bacteria that have undergone a favourable mutation that enables them to proliferate; the bacteria that have not undergone mutation are inhibited by the antibiotic and will be unable to grow. The role of the antibiotic in resistance development is to select mutants that are produced spontaneously, regardless of the presence of the antibiotic; it does not induce the mutation itself. This is a straightforward extrapolation from the Darwinian theory of evolution.

It is a very easy phenomenon to show in the laboratory. If you take any bacterium sensitive to an antibiotic like ampicillin, it may be sensitive to a concentration of 1 mg per litre; however, it may be able to grow in a concentration of ampicillin at 0.5 mg per litre, and we can then state that the minimum inhibitory concentration is 1 mg per litre. This MIC, as it is often referred to, is a good standard to compare bacterial sensitivities to antibiotics. If the bacterium, during the course of culture, undergoes a mutation

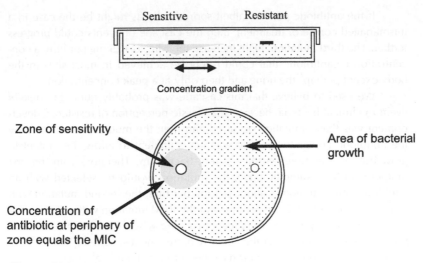

Disc sensitivity test

that provides relative insensitivity to ampicillin, then the presence of ampicillin in the environment, either by treatment or in the laboratory agar, will promote the growth of the mutant. When the mutant is isolated, the level of insensitivity to ampicillin can be determined simply by placing it on a series of agar plates, each usually containing twice the concentration of antibiotic of the previous plate. The new MIC can be calculated and typically if the MIC had been 1 mg per litre then that of the mutant may have risen to 5 mg per litre. This is quite a modest increase and if the antibiotic course was given conscientiously, a mutation giving an increase in MIC of fivefold should not present a particular problem. If, however, that mutant does succeed and the selection environment allows it to predominate, a cascade of events may have been precipitated. Instead of a fully sensitive population, there is now a partially sensitive collection of bacteria and, as these divide and proliferate, they also undergo mutations. If the selective environment is still present, these bacteria will succeed over their more sensitive parent bacteria. It used to be believed that the selective environment would have to be "stepped up" to allow these second mutations to be favoured. In fact, the original selection environment might be sufficient to favour the second mutations because although it may be at a concentration well below the MIC caused by the first mutation, the secondary mutant should still grow more quickly than the first mutant. The effect of the secondary mutation may be to raise the MIC from 5 mg per litre to 25 mg per litre. This is a level of resistance that may trigger clinical failure.

If the antibiotic persists, albeit sporadically as might be the case in a mismanaged course of treatment, then the cascade of events could progress further. The third mutation might increase the MIC to 125 mg per litre, a concentration of antibiotic that could rarely be achieved in most sites in the body except perhaps the urine and then only at a peak concentration.

We used to believe that this cascade was probably quite an unlikely event in clinical bacteria, because the classic perception of resistance development was that a mutation occurred and then the mutant was selected by the antibiotic concentration that was just sufficient to allow the mutant to grow but not the more sensitive parent bacterium. Therefore with the first mutation in the example above, the mutation would be selected with an antibiotic concentration of about 1–3 mg per litre, the second mutation with a concentration of 5–15 mg per litre and the third mutation with a concentration of 25–75 mg per litre. The progressive step-wise increases in concentration are most unlikely under the majority of treatment schedules, but bacteria with multiple resistant mutations have been found in clinical practice.

Our perception of how resistance progresses has been strongly influenced by our laboratory experiments to simulate resistance development. If we look at the same example and simulate a poorly administered course of antibiotic treatment, we can observe the same progression. It may be remembered that our sensitive hypothetical bacterium started with an MIC of 1 mg per litre; if the patient is treated with sub-inhibitory concentrations the following progression is likely. An environment of 0.5 mg per litre will allow all the sensitive bacteria to divide, albeit rather more slowly than they would in a drug-free medium. Approximately every 10 million replications of the bacterial chromosomal DNA, a mutation will occur to provide the bac-

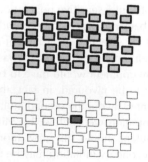

**Spontaneous Mutation
- Usually Independent of
Antibiotic Usage**

**Selection of Mutation
- Often by the Killing of
Antibiotic Sensitive
Bacteria**

Chromosome mutation

terium with an MIC of 5 mg per litre; this is the first mutation observed as before. As this mutant is less susceptible to the action of ampicillin, it will divide more rapidly than the parent sensitive strain. As the infection persists, the bacteria containing the first mutant will completely outgrow the sensitive bacteria and the infection is now caused solely by the mutant. The effect of sub-inhibitory concentrations on growth and the influence of a mutation may be seen in the figure on page 84.

As the mutant goes through successive cell divisions, approximately 1 in 10 million will produce the second mutation which will raise the MIC to 25 mg per litre. The antibiotic environment is still 0.5 mg per litre, but the second mutation is retarded even less by this concentration than the first mutant. The difference in the division rates may be less between the two mutants than between the first mutant and the sensitive parent, but it is still sufficiently different for the second mutant to outgrow the first. The progression can, of course, continue to produce multiple mutants with high degrees of insusceptibility to the antibiotic used for treatment. It is impossible to stress too strongly that this cascade of events is produced not by high concentrations of drug, but rather by insufficient and inadequate dosing. This is a theme to which we shall return time and again in antibiotic resistance emergence.

How could the initial cascade be prevented? If enough drug is administered so that the concentration of antibiotic at the site of infection is sufficiently high above the MIC, then bacterial division will cease and no mutants can be generated. If the infection comprises more than about one

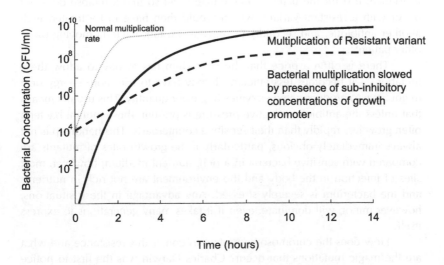

Effect of sub-inhibitory growth promoters at selecting resistant bacteria

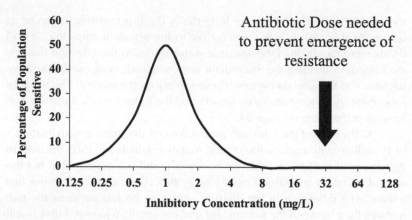

Why high levels of antibiotic are needed to prevent the emergence of resistance

million bacterial cells then, of course, it is possible that a mutant has already been generated and treatment has to take this into consideration. The basic rule of thumb is that the concentration at the site of infection should be at least four times the MIC and this should be maintained for as long as possible during the course of treatment. Antibiotics are given in fixed numbers of doses, perhaps 20 tablets, two taken twice daily for five days. Almost everyone is aware that patients are urged to complete their course, even if the symptoms disappear. This precaution is merely to prevent the emergence of resistance, first for the patient's immediate relief so that a relapse does not occur with a resistant variant, which would then have to be treated with another antibiotic, and secondly to prevent resistant bacteria moving on to other patients.

There is often a price that the bacterium has to pay to allow these mutations – it becomes less efficient. It may have to use more energy or it might have to produce some enzymes in greater quantity. This usually means that unless the antibiotic selective pressure is present, these mutant bacteria often grow less rapidly than their sensitive counterparts. This handicap is not always immediately obvious, particularly if the growth rates of mutants are compared with sensitive bacteria in a rich, nutrient medium; however, most sites of infection in the body and the environment are not rich in nutrients and the bacterium is severely stressed. Any advantage in these situations, however minor, will dominate, even if it takes many generations to express itself.

How does the chromosome mutate to confer this resistance and what are the magic mutations that occur? Charles Darwin was the first to notice that species varied in different environments; adaptations appeared that conferred an advantage over the others. This "survival of the fittest" theory was based on his observations of 14 slightly different finches in the Galapagos Islands, where he noticed particularly that the finches on each of the islands had evolved different-shaped beaks depending on the predominant food source; for the finch it was a case of mutate and adapt or die. Darwin believed that this fundamental principle was applicable to all species; it even applies to the bacteria that he knew nothing about. The only difference between bacteria and most other species was that they divide much more quickly, so they produce more mutations and these mutations will be manifested much more quickly. If a bacterium can divide once every 30 minutes, one bacterium will spawn 16,777,216 progeny in 12 hours; one finch would take nearly 100 years to produce the same number of descendants. In bacteria, evolution occurs in the fast lane.

It seems very abstract to describe one mutation in any gene that might confer resistance, but it just means that a mutation occurs in the one gene that could provide the bacterium with a selective advantage. I often ask my students, if they were designing a resistance mechanism against an antibiotic, what would it be. The most obvious answer seems to be some mechanism that prevents the antibiotic getting into the bacterial cell. We have seen that most antibiotics act by inhibiting a target inside the bacterial cell. If the antibiotic can be prevented from reaching the target then surely resistance must follow. Of course it does but it is only a matter of degree; if there is a massive challenge of antibiotic surrounding the bacterial cells, even if there is a mutation that restricts the passage into the cell, the sheer mass of drug trying to get in will often overwhelm any attempt to restrict entry. Mutations that restrict entry and thus confer impermeability are often inefficient and do not provide significant levels of resistance. They may be manifested by alterations in the proteins that surround the porin channels. The porins are passages through the outer membrane of the cell; these pores allow polar or water-soluble nutrients into the cell and polar antibiotics exploit them to gain entry to the cell. Closing the pores makes the bacterium less permeable. Other antibiotics, such as tetracycline, are molecules that are so large that they have to use a transport system to carry them into the bacteria. Impermeability is manifested in mutations that disable the enzymes of this transport system. Impermeability is, however, not a common mechanism of resistance because besides its inherent incapability to confer high levels of resistance, it also usually confers an enormous burden on the cell. Restricted passage of antibiotics through the porins will also impede the passage of vital nutrients.

The alternative would be to prevent the binding of the antibiotic to the target. Antibiotics have evolved to bind specifically to certain proteins and thus any alteration in the binding site of the target may reduce the capability of the antibiotic to attach. A mutation in the gene encoding the target binding site is the most common type of mutation. Its efficiency depends very much on the binding site itself. An alteration in the binding site of trimethoprim, the bacterial enzyme dihydrofolate reductase, confers only a low level of resistance; it might raise the MIC by only 50-fold. An alteration in the penicillin binding proteins, the targets for all the ß-lactam antibiotics, may increase the MIC by the same order of magnitude, for example methicillin resistance in methicillin resistant *Staphylococcus aureus*.

In clinical bacteria currently causing problems in hospitals, one of the most problematic manifestations of this type of resistance mechanism is resistance to the fluorinated quinolones. As we have seen, the problem with assessing the drugs is that the exact target of these drugs is unclear, thus the contribution of individual chromosomal mutations is difficult to assess. The traditional target of the drugs is the α subunit of bacterial topoisomerase II, popularly known as DNA gyrase, and alterations in this enzyme are associated with changes in resistance level. Changes at specific amino acids, most usually at positions 83 and 87 in the α subunit, are associated with resistance; however, the levels of resistance that they confer are often very variable. We have found that both mutations can be associated with increases in MICs to only 0.5 mg per litre in *Salmonella typhi* but are associated with much higher levels of resistance in *E. coli*. In these two relatively similar bacteria, surely these mutations cannot be responsible for vastly different levels of resistance? The answer is probably not; there must be other resistance mechanisms working *in concert*, the most likely one being alteration in a secondary target. Fluoroquinolones also bind to topoisomerase IV, encoded by the *parC* gene, and mutations can themselves cause increases in resistance. In the presence of mutations in the DNA gyrase subunit, the combination of resistance mutations can confer high levels of resistance. Which mutations come first is not really known yet. However, the chromosomal mutations causing alterations in these two targets are very successful in conferring resistance to the fluoroquinolones and significant levels are found in some pathogenic species.

Some alterations in the target can confer very high levels of resistance. The aminoglycosides bind to the protein S12 of the bacterial ribsome, leading to a distortion of the ribosome itself. The mutation causes a single amino acid replacement, substituting one or other of its two lysine moieties, which leads to a massive increase in resistance. This single mutation can raise the MIC by 2000-fold so the bacteria become totally resistant. This distortion of the ribosome can lead to a very interesting and strange capability: some bacteria have mutated so that they naturally have distorted ribosomes. They cannot, of course, grow in this state; however, the presence of an aminoglycoside binding to the protein contorts the distorted protein back to its active shape. Bacterial function is contingent on the presence of the aminoglycoside, i.e. the bacteria is aminoglycoside-dependent; it is, of course, highly resistant as well.

Although these levels of resistance will confer an advantage on the host bacterium in an environment with an antibiotic, it is not a permanent solution because these mechanisms confer an inherent energy drain on the cell. These mutations are usually not competitive if there is no antibiotic in the immediate vicinity, and revertant mutations are usually favoured when the antibiotic has dissipated.

Many clinical bacteria have been residents of soil at one time or another in their evolution. This has provided them with the defence machinery to deal with some of the natural antibiotics. The most prevalent is the protection against the β-lactam antibiotics, the penicillins, cephalosporins and carbapenems. The mechanism of aegis that evolution has favoured for most bacteria is the production of a β-lactamase. This is an enzyme that mimics the structure of the final target of β-lactams, the PBPs, so that this group binds tightly. The integrity of this antibiotic group is dependent on the four-membered planar β-lactam ring, which gives these drugs their name. If the structure of this ring is lost, the antibiotic is not active. Bacteria have found that the most efficient mechanism to overcome attack by these molecules is to destroy them before they reach the PBPs. The β-lactamase that they produce breaks the bond between the carbon and hydrogen of the β-lactam ring; then hydrolyse by adding the components of water. This means that the molecule loses its planar configuration, the whole molecule can swivel around the remaining bonds (see figure below), and it is no longer recognised by the PBPs and thus has lost any activity. This is a phenomenally successful resistance mechanism and there are over 260 β-lactamases found in clinical bacteria that can perform this one reaction. What is even more surprising is that the necessity to produce a β-lactamase is so great that they have evolved by at least four different routes. There are four distinct and completely different β-lactamase structures which have evolved independently. These β-lactamase classes have conveniently been called A to D. Three of these β-lactamase classes (A, C and D) have a serine amino acid residue at their active site involved in the catalysis of the acylation of the

β-lactamase

Active of a beta-lactamase enzyme, showing the less of the planarity of the beta-lactam ring

β-lactam ring. The final class (B) has a metal ion, usually zinc, at the centre of the active site. Almost all bacteria have their characteristic species-specific β-lactamase. In gram-positive bacteria these are usually from class A and have predominant activity against penicillins; in gram-negative bacteria they are usually from class C and have predominant activity against the cephalosporins. This raises the quandary as to how bacteria carry this excess baggage and why are they not inherently resistant to the β-lactams. Bacteria have had to evolve a system whereby they can call upon this β-lactamase when they need it but not expend unnecessary energy producing the enzyme when they do not require it. They achieve this largely by switching off enzyme production for the time that it is not needed. They produce a protein, called a repressor, that binds and inhibits to the promoter region that initiates the decoding of the β-lactamase gene. So β-lactamase is not produced. When the bacteria enter an environment where there are β-lactam molecules trying to enter the cell (in the soil this might be on contact with a fungus), the bacteria have a signalling system that transmits the message of this attack inside the cell. The harbinger molecule is preferentially bound by the repressor protein which, because it is now engaged, cannot bind to the promoter of the β-lactamase gene and, in the absence of this repression, the β-lactamase is produced in maximal quantities. The β-lactamase is transported to the outside of the cell. In gram-positive bacteria it is usually exported out of the cell into the surrounding medium and attempts to destroy all the β-lactam molecules in the immediate vicinity of the cell, whereas in gram-negative bacteria it is exported to the periplasmic space between the inner and outer membrane to intercept the incoming antibiotic attack. So bacteria are sensitive to β-lactams unless they can induce the β-lactamase, which occurs when the bacteria meet an antibiotic challenge.

This is a finely tuned defence system to deal with the war of attrition between the various armies of microorganisms competing for nutrients in the soil. The antibiotic weapons are released slowly and attack is protracted. Antibiotic use in clinical practice is *blitzkrieg*, carpet bombing which requires instant retaliation; however, the induction of the β-lactamase is simply too slow because, by the time the enzyme is produced in sufficient quantities, the host bacterium has succumbed to the attack. The only hope that the unprepared population of bacteria could have in this predicament is that, while most cells are killed by the attack, some might have time to induce the β-lactamase and the population will be saved by the progeny of these induced producers. In fact, this is often not the case and induction is insufficient protection. To deal with this new challenge, bacteria have evolved to continuously produce the β-lactamase at maximal amounts. They achieve this by mutation in the gene that encoded the repressor; the energy expenditure required to service the consequences of the permanent removal of the restriction on the promoter of the β-lactamase gene would not normally be tolerated in these bacteria but their survival depends on it. In some bacterial species, such as *Enterobacter* and *Serratia*, the pressure to overcome β-lactam attack is so great that up to 60% of all clinical isolates carry this mutation. This has been called stable de-repression but should perhaps strictly be more correctly termed constitutive production of β-lactamase; the mechanism of resistance is not mutation in a structural gene to cause resistance but rather a mutation in a control gene to increase the number of defence molecules.

This resistance mechanism can be termed hyperproduction and is not just confined to β-lactams. Chromosomal dihydrofolate reductase, the target of trimethoprim, can be hyperproduced in some cases. This simply steps up production to increase the number of available dihydrofolate reductase active sites. If, say, 10 active sites are required in the cell to reduce dihydrofolate to tetrahydrofolate then when trimethoprim tries to bind and inhibit these active sites, increased production of enzyme will produce more active sites, some of which will be inhibited by the antibiotic, but as long as sufficient are produced so that 10 remain uninhibited for the reduction of dihydrofolate, the action of the drug is checked.

5

Sharing knowledge – the spread of antibiotic resistance

PLASMIDS

During the 1950s, resistance emerged to almost all antibiotics in use. It became predictable that after 10 million bacterial cell divisions, a mutation would occur. This was easy to demonstrate not only in the laboratory but also in the clinic. It meant, however, that if there was a potential problem with resistance, the clinician could give two drugs. The chances of two mutations to confer resistance to two drugs is 10 million x 10 million, a figure so large that we can only express it as a power of 10, 10^{14}. Actually this number of bacteria represents a volume of nearly a one litre solid mass of bacterial cells, which even with the least pathogenic bacteria would be more than sufficient to kill the patient. Indeed it would be difficult to locate the infected organ within this mass of bacteria. The point is that this figure simply shows that too many cells are required for a double mutation to manifest itself. Of course, if the bacterium was already resistant to one of the antibiotics, and it was challenged with two drugs together, then the mutation rate would be 1 in 10 million because only one mutation would be required to render the bacterium resistant to both drugs. But dual therapy presupposes that the bacterium is sensitive to both drugs.

With the preconception that mutation to resistance occurred at a constant rate, Japanese microbiologists were astonished to find, in 1959, that some *Shigella* strains, responsible for dysentery, were apparently mutating to resistance to four antibiotics at the same time. Under the contemporary understanding of resistance development, this would involve a mutation in the order of 1 in 10^{28}, a mass of bacteria that could easily engulf the whole planet; clearly some other mechanism had to be operating. The Japanese microbiologists, Akiba and Ochai, examined their bacteria again to ensure that there was no build-up of resistance, but there were simply no bacteria resistant to one, two or three antibiotics; they were either resistant or sensitive to all four antibiotics. The clue came when they made the remarkable discovery that the bacteria were able to transfer these resistances from one bacterium to another. The resistance characters transferred *en bloc*; what they had found was that genes for resistance were not mutated genes located on the bacterial chromosome but rather they were present on a separate molecule of DNA. This independent DNA molecule had the capability to replicate independently of the bacterial chromosomal DNA and, more astonishingly, could transfer a copy of itself from one species to another. The original Japanese observation had been this transfer. They published their results in *Nihon Iji Shimpo* in Japanese, so these results remained hidden from the scientific community, which is almost exclusively English-speaking, for four years. When Watanabe published a review on resistance in 1963, microbiologists were astounded, for this explanation did account for many results that had been observed by other workers at that time. Finding a name for this independent transferable DNA molecule was difficult; it was called an episome, a term designed to encompass all independent DNA elements, including those that had the capability to integrate into the bacterial chromosome. An alternative term was plasmid, but these elements were considered strictly independent of the chromosome. It was impossible to guarantee that any DNA could remain independent of the chromosome under all conditions, so plasmid status was impossible. However, the name was favoured and no mention is ever made now of episome, regardless of whether the DNA element is independent of the chromosome; they are all plasmids.

The nature of the resistance genes themselves was an intriguing problem for biochemists; were they simply chromosomal resistance genes now located on plasmids and, if so, why did they not just stay on the chromosome? Close analysis revealed that these early plasmid-encoded resistance genes were quite unlike their chromosomal counterparts and they did not have a common origin. Ampicillin had been released at approximately

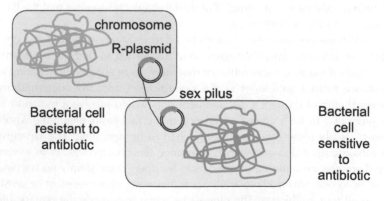

Bacteria acquire new resistance genes by plasmid transfer

the same time as plasmids were first discovered. Plasmid resistance to ampicillin, usually characterised by its ability to transfer, was suddenly very common. Most common pathogenic bacterial species were showing some infiltration by ampicillin resistance. β-lactam resistance had originally been shown to be chromosomal and it was not clear how it had moved to plasmid. The resistance, like that encoded by the chromosome, was manifested by β-lactamase but they were not the same enzymes as any found encoded by bacteria themselves. Their production was not repressed as it had been with chromosomal β-lactamases. These enzymes were not indigenous to the bacteria that they were now inhabiting, but had been imported from quite distant species.

We have seen how virtually every bacterial chromosome encodes a β-lactamase and that these may be one of four major classes, each one unrelated to the others. Members of each class have been found to be encoded by the chromosome of at least one bacterium, though some are much more common on plasmids. The class D β-lactamases are almost invariably found to be encoded by plasmids and until recently were considered to be only of plasmid origin. Recently, one of the chromosomal β-lactamases of *Aeromonas* species has been found to have a class D structure and, in the absence of other chromosomal contenders, this might be considered the archetype of the class. However, amongst the plasmid-encoded β-lactamases in this class there is some variation in structure and a wide variation in their biochemical properties; for instance the so-called OXA sub-class of class D β-lactamases have proved efficient at hydrolysing the isoxazolylpenicillins and methicillin whereas the PSE sub-class are effective against carbenicillin. The origin of these β-lactamase will not be clear until more chromosomally class D β-lactamases are found.

Seven years ago a very astute PhD student of mine, David Payne, was examining an isolate from a patient referred to a London hospital from Pakistan. Cephalosporins would normally be effective against the *E. coli* that continuously infected the patient, but of the myriad available none could eradicate the infection. David Payne's examination of the biochemical properties of the β-lactamase, which we called BIL-1, suggested that it was an example of a β-lactamase normally associated with the chromosome of gram-negative bacteria, the class C enzymes. He showed that the resistance was transferable to other *E. coli* and proposed that the gene must be located on a plasmid. He determined the sequence of the gene and revealed that it was derived directly from the chromosomal β-lactamase of *Citrobacter freundii*. This was one of the first and still best examples of the direct emergence of plasmid-encoded resistance derived from chromosomal genes. Similarly the MIR-1 β-lactamase, identified in the United States, has similar biochemical properties to BIL-1 and partial sequencing of the gene shows that it is probably derived from *Enterobacter cloacae*. When David Payne

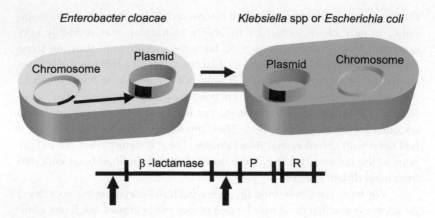

Enterobacter cloacae P99 β-lactamase genes

and I first made the original observations, we probably happened across the acquisition of Class C β-lactamase genes by plasmids as it was first emerging. The function of these genes in plasmid vectors is rather different from that in the chromosome of the original bacterium; there they had been repressed, the restrained guardian waiting for the foe, but relocation in a plasmid ensures that the gene is not repressed and is constitutively expressed. Why should these two β-lactamase genes migrate to the plasmid? There may be two reasons, the first that it permits them to spread from one bacterial cell to another. If it is assumed that genes are selfish, then the capability to disseminate to other bacterial species must be advantageous because it increases the capability to survive. It is also likely that plasmid carriage satisfies another need, that of experimentation. Both the BIL-1 and MIR-1 β-lactamases are clear adaptations of chromosomal genes, but they are not the original genes. The MIR-1 β-lactamase from Boston has a substrate profile most similar to *E. coli ampC*, particularly in its capability to hydrolyse cephalothin. In this respect, it is dissimilar to the chromosomal β-lactamase of *Enterobacter cloacae,* a relation from which it may derive. Thus when the β-lactamase gene was extracted out of some distant *Enterobacter* spp onto a plasmid, it had to adapt under different selective pressures – perhaps under cephalothin therapy. So the plasmid carriage of the resistance gene may have allowed mutation and variation without damage to the host bacterium. If mutations occur in a chromosomal gene, they might be lethal and the cell might die. Lethal mutations in plasmid genes are far less damaging; after all, there will be other plasmid-carrying cells in the population that have not undergone the mutation.

As defence mechanisms, the plasmid-encoded class C β-lactamases have the additional advantage that they are not susceptible to the common β-lactamase inhibitors, clavulanic acid, sulbactam and tazobactam. These inhibitors are produced by the *Streptomyces* spp and presumably evolved to inhibit the β-lactamases of competitive bacteria, particularly the class A chromosomal β-lactamases of gram-positive bacteria. In a penicillin-rich environment, all β-lactamase-producing micro-organisms will survive but those that are capable of inhibiting the β-lactamases of competitors will have an advantage.

The most studied and the largest group of plasmid-encoded β-lactamases is class A. Like their gram-positive chromosomal progenitors, they all possess an active serine molecule at position 70 and probably originally derived from the genes involved in the final stage of peptidoglycan synthesis, probably the *Streptomyces* R61 DD-peptidase. All class A β-lactamases have significant conservation of the amino acid residues in critical areas of the sequence, which ensures that the essential 3-D shape of the enzyme is maintained. Huletsky and colleagues recently conducted a close comparison of the structures of the genes encoding the class A β-lactamases and produced a dendrogram which diagramatically showed the phylogenetic relationships between them. They suggest that the chromosomal β-lactamases of gram-positive bacteria, particularly *Streptomyces*, *Staphylococcus* and *Bacillus*, emerged early in evolution. It might be that once these bacteria were facing β-lactam attack from soil fungi, that they may have tried to alleviate the onslaught by increasing the number of active targets to soak up the intruder. The increase in targets might have resulted from duplication of the DD-peptidase gene used in the final synthesis of the cell wall. Once the gene has duplicated, mutations can occur in one of the gene copies without causing irreparable damage to the cell. If, for instance, the mutation results in an

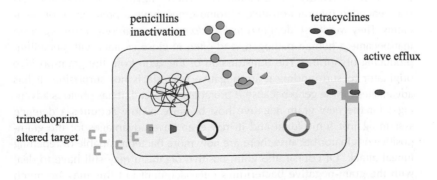

Mechanisms of antibiotic resistance

inactive DD-peptidase, the bacterium still has the unmutated DD-peptidase gene to fall back to. If, on the other hand, the mutation not only causes the new protein to bind the β-lactam but also it can hydrolyse the carbon–nitrogen bond of the β-lactam ring, then the host strain has an enormous advantage. The emergence of the β-lactamase did not seem to derive by a spontaneous mutation; it was probably preceded by gene duplication which would minimise damage caused by mutation and then a sequence of mutations, each conferring a slightly greater advantage than the previous one, might occur to increase the binding capability of the DD-peptidase. When a sufficient number of mutations have occurred this molecule could now not only bind but hydrolyse and release the disabled β-lactam.

Gram-positive bacteria are simpler than gram-negative and it assumed that they evolved first. It is known that the genes can readily transfer from gram-positive to gram-negative bacteria but not the other way round. Most of the "early" β-lactamases are considered to be the chromosomal enzymes of gram-positive bacteria except, perhaps, the plasmid-encoded PC1 β-lactamase of *Staphylococcus aureus*. Although there are some notable differences between the β-lactamases of these gram-positive bacteria, which might reflect the aeons involved in their evolution, in molecular terms they are quite similar and reflect a common origin. At one time, maybe even quite recently, the PC1 β-lactamase gene migrated to a plasmid. There is one plasmid-encoded class A β-lactamase found only in gram-negative bacteria, ROB-1, which resembles these gram-positive enzymes. It is found only in *Haemophilus influenzae*, a respiratory pathogen, and its structure suggests that its gene has recently been transported from a gram-positive species into an *Haemophilus* plasmid.

The first plasmid-encoded class A β-lactamases that appeared in gram-negative bacteria are likely to be those associated with *Pseudomonas aeruginosa*, the so-called PSE (Pseudomonas specific enzymes) and CARB (Carbenicillin hydrolysing) β-lactamases. These enzymes, like their counterparts encoded by gram-positive chromosomes, have a preference for penicillins. They were first identified in *Pseudomonas aeruginosa*, a soil organism that became a human pathogen, and often in close contact with penicillin-producing organisms. The transmission of the gene from the gram-positive originator to surrounding gram-negative bacteria is not surprising. It has advantages for the gene because it promotes its spread, it has obvious advantages for the new gram-negative host because it now acquires a defence system against some fungi and it may even have advantages for the gram-positive originator because there are now more bacteria capable of repelling fungal attack. Of course, the gram-positive organism may still have to deal with the gram-negative bacterium's encroachment but this may be much easier than fungal attack.

The most prolific and most worrying plasmid-encoded class A β-lactamases are the TEM and SHV enzymes. The SHV β-lactamases are similar to some chromosomally encoded β-lactamases in *Klebsiella* spp. It is likely that the genes for the SHV enzymes were donated to the *Klebsiella* spp in the soil environment from a passing gram-positive organism. We shall see shortly how this gene might become a permanent resident of the chromosome of the gram-negative bacterium; from there it is easy to donate a copy to plasmids.

The TEM family of class A β-lactamases is, by far, the most common plasmid-encoded resistance gene in clinical gram-negative bacteria. They derive their name from the patient Temeida who carried the bacterium which carried a new resistance plasmid; it was called R-TEM. The plasmid was soon renamed by the less obvious label R6K but the β-lactamase it carried is still called TEM-1.

The plasmid-encoded TEM-1 β-lactamase represents one of the greatest impacts that Man has had on his environment. It is, by far, the most common resistance gene not just for β-lactam antibiotics but for all resistance genes. There are over 100 different plasmid-encoded β-lactamases identified in clinical bacteria and, in any survey of common pathogens, TEM-1 *always* represents a minimum of 75% of plasmid-encoded β-lactamases. So successful is this gene that it is found in bacteria of the normal gut flora in 25% of the healthy British population, people undergoing no antibiotic therapy. In India this figure is nearly 100%; this really is a super-gene.

It is impossible to identify a likely species of origin of TEM-1 β-lactamase, perhaps because it is so widespread. The TEM β-lactamase gene has about 65% identity with the SHV-1 gene but in terms of their biochemical properties the β-lactamases that the two genes encode are identical. Huletsky *et al.* suggest that both genes were derived from the same source but the close similarity in enzyme properties suggests that selective pressures for both enzymes have been much the same. Why then is the TEM-1 more common by a factor of nearly 100-fold? There are likely to be two major contributory factors; its success may derive either from the respective mobility of the genes or subtle differences in their biochemical properties which our relatively crude assay techniques are just too insensitive to discriminate. Mobility is a difficult parameter to quantify as both SHV-1 and TEM-1 are plasmid-encoded. They are often located on successful plasmid and also often present on transposons. It seems more likely that subtle differences in the ability to bind and hydrolyse β-lactams have favoured the TEM-1 enzyme; by far the most widely used antibiotic in the world has been ampicillin and its sibling compound amoxycillin. They are cheap and have been used in huge quantities for the past 30 years. The TEM-1 enzyme seems to be the β-lactamase that has evolved the active site that binds these two drugs most effectively.

The subtlety of the active site can be seen with the β-lactamase known as TEM-2. This enzyme is identical to TEM-1, except that it possesses a lysine residue at position 39 instead of glutamine. This substitution occurs at an amino acid residue which seems quite distant from the active site of the enzyme and has often been considered to have little influence on it. Despite this, the TEM-1 β-lactamase is 10 times more common than the TEM-2. However, any change in the structure might have an effect on the thermodynamic stability of the molecule, crucially on its exact 3-dimensional configuration and, in particular, with the concomitant effect on the spatial arrangements within the active site. The minute alterations to the active site, the effects of which we are incapable of measuring, may mean that the capability of TEM-1 to bind and hydrolyse to ampicillin and amoxycillin remains unsurpassed.

I suggested that plasmids have a role as a medium for rapid experimentation to create resistance genes to contend with new selective pressures without long-term damage to the host bacterium. At the present time, we are witnessing a Herculean struggle occurring in clinical bacteria to foster the evolution of plasmid-encoded genes rebutting the introduction of new antibiotics. The need to control the ubiquitous TEM-1 β-lactamase has mesmerised pharmaceutical companies so that they have striven to devise sophisticated cephalosporins that would not be hydrolysed by these enzymes. Their final triumph was what they have often described as the "third generation" of cephalosporins. We might epitomise these antibiotics by ceftazidime and cefotaxime. Both cephalosporins were considered to be supreme achievements in the design of drugs developed to neutralise a ubiquitous plasmid-encoded resistance mechanism. The structure of the active site of the TEM-1 β-lactamase suggested that these cephalosporins simply could never enter it; if these cephalosporins were unable to bind then they could not be hydrolysed. They still entered the outer membrane of the cell sufficiently well to bind to the PBPs but they do not enter as well as the penicillins, so they are often not as active. However, this is the compromise required to retain a β-lactam antibiotic which was effective against TEM β-lactamase-containing bacteria. It was believed, at the time, that a β-lactam drug had been conceived for which there would be no plasmid-encoded resistance.

Ceftazidime was a cephalosporin that had been devised because it was particularly effective against *Pseudomonas aeruginosa*; however, this pathogen is relatively rare in hospitals though therapeutic options often take its potential presence into account. Thus ceftazidime enjoyed widespread usage in hospitals almost immediately after its launch. Two years after its introduction, ceftazidime was being used to treat a *Klebsiella oxytoca* outbreak in a neonatal unit in Liverpool, England. Against the causative bacteria identified early in the outbreak, the drug worked effectively but subsequently it was found to be ineffective – the *K. oxytoca* had become resistant. It was not very resistant, with an MIC of around 4 mg per litre, but this was sufficient to ensure that successful eradication could not be guaranteed. Examination of the ceftazidime-sensitive bacteria at the start of the outbreak showed that the bacterium harboured a plasmid which encoded a TEM-1 β-lactamase, surely the very type of strain that should get ceftazidime treatment. Initial examination of the resistant isolates showed that the ceftazidime resistance could transfer from one strain to another and was presumably located on a plasmid. When the β-lactamase was examined in the transconjugant, it looked to be TEM-1 and was mis-classified. In fact the TEM-1 β-lactamase had mutated and the clinical use of ceftazidime was selecting bacteria containing this mutation. The plasmid carrying the TEM-1 gene remained completely unchanged during ceftazidime treatment but a single nucleotide, within the TEM-1 β-lactamase gene, was altered, changing the amino acid at position 164 from arginine to serine. Arginine is a basic amino acid and can form ionic bonds with acidic amino acid residues. There are two in the vicinity, glutamic acid at position 171 and asparatic acid at 179. Arginine forms ionic bonds with both of them and, in doing so, brings the small α-helix formed by the amino acid sequence between 164 and 171 to the entrance of the active site. In fact, it partially blocks the entrance and this is one of the reasons why the active site is so well suited to the binding of ampicillin/amoxycillin, and cephalosporins are simply too large to pass this sentinel. The mutation to substitute serine at position 164 does not rely on any special property of serine, rather it is the loss of arginine and the concomitant capability to form the ionic bonds that is the crucial factor. If the ionic bonds are broken, the small α-helix falls away from the entrance to the active site and the larger cephalosporins can now bind. Although they can now bind quite well, they are not very efficiently hydrolysed; however both ceftazidime and cefotaxime are hydrolysed to approximately the same extent. This unit hydrolysis confers a level of resistance to ceftazidime that is on the borderline between clinical resistance and sensitivity; it gives an MIC of around 4 mg per litre. Although of limited clinical importance, it assumes a much more important role when it is seen as the start of a cascade. This selection of resistant mutants is very reminiscent of the classic mutation and selection of antibiotic-resistant chromosomal genes in clinical bacteria and is unlike the earlier response employed by plasmids of importing genes from outside. Very significantly, although the hydrolysis rate is equivalent to ceftazidime, this mutation confers no resistance to cefotaxime. Cefotaxime was a cephalosporin designed for use against the Enterobacteriaceae; it penetrates the porins rapidly and moves quickly to the PBPs. Ceftazidime, the antipseudomonas cephalosporin, penetrates slowly; thus for the same measure of β-lactamase activity, the mutant enzyme is much more efficient at conferring resistance to ceftazidime. This differential is absolutely crucial because the 164 mutation is the pivotal mutation of most TEM-derived extended-spec-

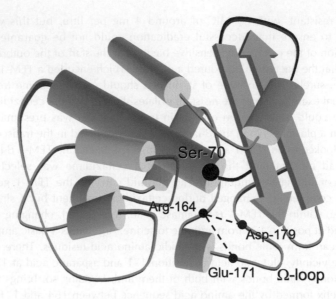

The key mutation in extended-spectrum activity for the TEM β-lactamase

trum β-lactamases and it appears that it may be selected only by slow-pene-trating cephalosporins, such as ceftazidime.

The β-lactamase caused by this mutation is known variously as TEM-E2 or TEM-101 but is generally commonly accepted as TEM-12. The parallel with chromosome mutation continues as more mutants are selected. Under further clinical treatment, the TEM-12 β-lactamase gene can mutate again to alter the amino acid at position 240 from glutamic acid to lysine. These amino acids are located in the β-sheet 3 on the right of the active site; the substitution pulls this sheet away from the active site and this opening of the active site further promotes the binding of ceftazidime. It causes a substantial increase in ceftazidime hydrolysis and a moderate increase in cefotaxime binding but still only confers ceftazidime resistance; again this results from the meagre penetration of ceftazidime. In fact, there is a substantial increase in the MIC of the host bacterium to the clinically significant level (MIC @ 64 mg per litre). This enzyme, with its double mutation, has been known as TEM-E3 but is more commonly called TEM-10. It has been the cause of significant ceftazidime resistance in gram-negative bacteria in many parts of the world, not least in many hospitals in London. The selection of the TEM-10 β-lactamase in the laboratory can only usually be achieved in the pres-ence of ceftazidime and not with a fast-penetrating cephalosporin. There is no direct clinical evidence of ceftazidime selection of this enzyme. The con-current acquisition of cefotaxime resistance was first found in a clinical strain identified in France. The strain carried a TEM enzyme which was identical to TEM-10 except that the amino acid at position 237, which had been alanine, was now threonine. This amino acid substitution also occurred in the β-3 sheet. The mutation is thought to pull the β-3 further away from the active site. There is no direct evidence to demonstrate that cefotaxime selected a mutant of TEM-10 in the clinical strain; however, the selection of the mutant can only occur in the laboratory if a fast-penetrating cephalosporin, such as cefotaxime, is used as the challenging selection agent in *Klebsiella* strains harbouring TEM-10. This triple mutant, known as TEM-5, has markedly increased hydrolysis of cefotaxime over ceftazidime, but still the resistance conferred to ceftazidime is higher than to cefotaxime, further demonstrating that the speed at which the cephalsoporin enters the porins is a crucial factor.

If *Klebsiella* strains are treated with sub-limiting concentrations of cefotaxime, we obtain mutations in the amino acid at 238, also located on the β-3 sheet. There are actually very few clinical mutants that have a substitution of glycine with serine at position 238; this has a similar effect as the mutation at 237 by pulling the β-sheet away, but this mutation does not need any previous mutations to manifest itself so it alone can allow cefotaxime to bind readily and thus endow sufficient resistance to cefotaxime on the host to be clinically important. This single mutant is called TEM-19 and still confers higher ceftazidime resistance than cefotaxime, though it is unlikely that ceftazidime has selected it. Its rarity in clinical bacteria suggests that, in the TEM-1 nucleus, this is an unstable mutation and will probably only persist in

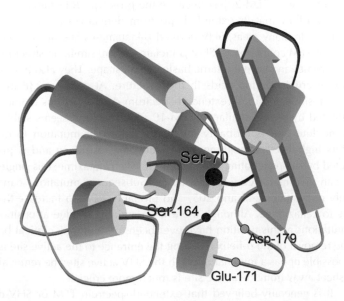

Opening of the active site of the TEM β-lactamase for extended-spectrum activity

an environment of continuous challenge with cefotaxime or similar fast-penetrating cephalosporin.

The only other important mutation site in the TEM-1 nucleus is the substitution of lysine for glutamic acid at position 104, on the left of the active site. The removal of acidic glutamic acid in favour of the basic lysine is thought to promote binding of cephalosporins, particularly ceftazidime. In its phenotype, this mutation is very similar to the 240 substitution on the other side of the active site; its effect is only apparent if arginine has previously been substituted at position 164.

A surprising observation about the extended-spectrum β-lactamases is that approximately half of those found in clinical bacteria are derived from the TEM-2 β-lactamase rather than the TEM-1, which is astonishing cognisant of the fact that the protype TEM-1 outnumbers TEM-2 by 10-fold in clinical bacteria. The preferential capability of TEM-2 β-lactamase to spawn so many extended-spectrum β-lactamases may derive from a peculiar side-effect of the apparently silent mutation at position 39; Baquero and colleagues found bacteria harbouring the TEM-2 β-lactamase seem to survive cefotaxime challenge better than strains harbouring TEM-1, so the distant substitution can affect the integrity of the active site. If TEM-2-containing strains survive longer under cefotaxime challenge, they have a greater chance to manifest extended-spectrum mutations.

The first extended-spectrum β-lactamases to be investigated were not derived from the TEM β-lactamases but from the far rarer SHV-1 enzyme. However, as with TEM-2, just because the prototype β-lactamase is rare, it does not follow that its extended-spectrum derivatives are proportionally infrequent. In many areas, SHV-derived β-lactamases are far more common than their TEM counterparts. SHV β-lactamases are similar in structure to the TEM enzymes and have the same basic overall shape. They also have conservation at crucial amino acids in the structure. At position 238 there is a glycine residue and in all extended-spectrum derivatives this amino acid is substituted by serine, similar to TEM-19. This mutation seems stable in the SHV nucleus and is probably the first and pivotal mutation to occur. It confers high-level ceftazidime and cefotaxime resistance and is probably selected by fast-penetrating cephalosporins like cefotaxime; this single mutation can give major clinical resistance. Subsequent mutations can occur, mainly at positions 240 and 205, and probably serve to increase the resistance to ceftazidime. Although there is an arginine residue at position 164, no mutations at this position have ever been found in any clinical bacteria, so the removal of the α-helix blocking the entrance to the active site is either not possible or has a minimal effect in the SHV active site; the removal of the β-3 sheet away from the active site is much more crucial.

It is generally believed that extended-spectrum TEM or SHV-derived β-lactamases are derived solely from successive challenges during the clinical use of cephalosporins but the selective pressures on clinical bacteria are not so simple. It is convenient to suggest that the evolving β-lactamases are becoming more and more resistant to cephalosporins as they encounter either higher doses or even newer versions of these drugs. This hypothesis ignores one crucial influence on these bacteria: even with the massive increase in cephalosporin use, the major β-lactam selection pressure on clinical bacteria is still ampicillin or amoxycillin, whether this is used alone or in combination with a β-lactamase inhibitor. As the β-lactamase develops resistance to cephalosporins, what happens to its capability to hydrolyse ampicillin/amoxycillin and to bind β-lactamase inhibitors? In short, the extended-spectrum mutations affect this ability drastically; invariably the acquisition of improved cephalosporin activity is accompanied by a decrease in efficiency against ampicillin/amoxycillin. This might be expected because the parental enzymes, TEM-1, TEM-2 and SHV-1, have evolved to confer resistance to these penicillins and their active sites bind them tightly. Any increase in size of the active site to permit entry by the cephalosporins inevitably relaxes the binding of the penicillins. Examination of some TEM-derived β-lactamases shows that there are some enzymes (TEM-13, TEM-17, TEM-18) whose presence in clinical bacteria cannot be explained by forward mutation and selection with cephalosporins and must have involved some interactions with either these pencillins alone or in combination with a β-lactamase inhibitor.

The conflicting effects of these different forces can be seen at their most simple in the progression from TEM-1 to TEM-5. This is a series of three successive mutations which can be seen in the figure below. AS TEM-1 mutates first to TEM-12, then to TEM-10 and finally to TEM-5, the efficiency of the enzyme for cefotaxime and ceftazidime increases but there is a corresponding decrease in efficiency for ampicillin/amoxycillin. In theory, if bacteria harbouring the TEM-5 β-lactamase were treated with ampicillin/amoxycillin, back mutations to TEM-10 should be favoured by this environment. In fact, this is difficult to demonstrate in the laboratory; however, the progression from TEM-1 to TEM-5 confers one other property: it increases the binding of the β-lactamase-inhibitor clavulanic acid. TEM-5 is much

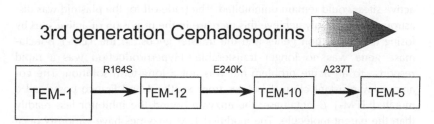

Sequential mutation of TEM β-lactamase to give extended spectrum activity

more readily inhibited than TEM-10, so if TEM-5 containing bacteria are now challenged with co-amoxiclav, the combination of a amoxycillin and clavulanic acid, reverse mutations that substitute alanine for threonine at position 237 to give TEM-10. TEM-10 both binds clavulanic acid less tightly and hydrolyses amoxycillin more efficiently than TEM-5, so this mutation will be favoured. Similarly if bacteria harbouring TEM-10 are challenged with co-amoxiclav, reverse mutations to TEM-12 are favoured. Eventually continuous challenge will result in selection of the parent enzyme TEM-1. The action of co-amoxiclav may not necessarily select a direct back mutation. The β-lactamase TEM-26 seems to result from a mutation in TEM-12 selected in the presence of high concentrations of slow-penetrating cephalosporins. Treatment of strains carrying TEM-26 might seem to favour back mutations to TEM-12 by a reversal of the glutamic acid to lysine substitution at position 104. However, if the substitution was a reversal of the arginine to serine substitution at position 164, the resultant β-lactamase, TEM-17, confers more resistance to co-amoxiclav than TEM-12. TEM-17 has no extended-spectrum activity; indeed the only advantage that can be ascribed to it is the increased resistance to co-amoxiclav over TEM-26. If the strain had emerged with TEM-17 and is challenged with cefotaxime treatment, it might allow a substitution at 238 to give TEM-15 and subsequently TEM-4. The extended-spectrum β-lactamases in clinical bacteria are undergoing a gigantic uncontrolled experiment in selection pressures and we have to speculate, largely in retrospect, what happened.

The end-product of co-amoxiclav challenge to extended-spectrum β-lactamases might appear to be the parental enzyme TEM-1. In 1992, there were a number of reports from Scotland and France that the TEM-1 β-lactamase had another capability, to undergo mutations, when carried by plasmids in *Escherichia coli,* to resistance to co-amoxiclav. Hyperproduction of the TEM-1 β-lactamase had already been shown to be a ready, if not efficient, mechanism of co-amoxiclav resistance. Sara Nandivada in my laboratory demonstrated that, in their drive to overcome the challenge of co-amoxiclav, plasmids underwent some tortuous rearrangements to increase the production of the β-lactamase and, therefore, the number of active sites that the deadly clavulanic acid could bind, in the anticipation that sufficient active sites would remain uninhibited. The trade-off for the plasmid was disastrous; it could only achieve this increase in the provision of active sites by losing the replication genes and the transfer genes, so the TEM-1 β-lactamase gene was no longer transferable. Hyperproduction was a rapid response to an acute problem but was not a long-term solution. The co-amoxiclav-resistant clinical isolates from Scotland and France possessed a mutated TEM-1 β-lactamase; the enzyme bound the inhibitor less readily than the parent molecule. The modified TEM enzymes have variously been called TRC (TEM Resistant to Clavulanic acid), TRI (TEM Resistant to

Inhibitor) or now the most common nomenclature IRT (Inhibitor Resistant TEM). In fact, at least 12 different inhibitor-resistant TEM β-lactamases have been described. The key mutation is usually identified as an alteration to arginine 244 in the β-4 sheet on the right side of the active pocket. This substitution, like that at 164 in the extended-spectrum β-lactamases, is with a shorter, uncharged residue such as serine (IRT-2), cysteine (IRT-1) or threonine (IRT-11). The water molecule involved in the binding of the inhibitor sets up an ionic bond with arginine, but when a smaller, uncharged residue is substituted, this bond cannot form. In fact, a more common mutation is at methionine-69, the adjacent amino acid to active serine-70. This residue lies at the back of the active site and is always substituted by an aliphatic, hydrophobic residue. This can be either isoleucine (IRT-3 and IRT-8), leucine (IRT-4, IRT-5 and IRT-10) or valine (IRT-6 and IRT-7). The presence of this non-polar residue changes the micro-environment at the active serine-70 residue, so that the inhibitors do not bind so readily. Other substitutions occur at in the arginine-275 and asparagine-276 residues, but they seem to affect the binding of the inhibitor to arginine-244, which in turn causes resistance. Interestingly, these mutations do not seem to affect the capability of the β-lactamase to hydrolyse ampicillin/amoxycillin and for this reason these mutants do seem different from the extended-spectrum mutations. They do not confer high levels of resistance but do appear to give the strain a selective advantage in a co-amoxiclav environment. In a recent study, 8% of all urinary *E. coli* in a French hospital were shown to possess these enzymes.

In the same way that co-amoxiclav can reverse the extended-spectrum mutations, cephalosporins do seem to select for reversals of the IRT mutations. Can both mutations occur in the same TEM molecule? The answer is yes and this has been found in clinical bacteria; however, the enzyme does not have both properties as it retains extended-spectrum capability but demonstrates no inhibitor-resistance even though the mutations is present. It might be assumed that the extended-spectrum mutation is dominant but subsequent isolates of mutant TEM β-lactamases may show us that this is not always true.

The most prolific source of antibiotics has been the inhibitors of protein synthesis; for some reason, inhibition of competitors' ability to produce viable proteins has been the most potent method to hinder the growth of other microorganisms competing for the same nutrients. The genus that has been the most abundant in providing antibiotics has been the *Streptomyces*, which have produced many antibiotics, particularly the aminoglycosides. However, these antibiotic-producing bacteria have a problem: they have to protect themselves against the assassins they manufacture. The defence mechanisms that evolved are quite unlike the chromosomally encoded antibiotic resistance mutations of clinical bacteria, where insusceptible targets emerge but although these are able to overcome the immediate danger from the antibiotic environment, they disable the ability of the bacterium to compete once the antibiotic has dissipated. In the highly competitive environment of the soil, crippling mutations would present a fatal disadvantage. This is a problem that confronted antibiotic-producing microorganisms such as *Streptomyces* long ago and they have had aeons to solve it. Clinical bacteria, on the other hand, have been challenged with antibiotics for only 50 years and they have not yet evolved the most efficient mechanisms of chromosomal resistance.

Streptomyces spp. could hardly be more different from the pathogenic bacteria where antibiotic resistance causes so many problems. Instead of following the normal growth and fission cycle of most clinical bacteria, they form elongated, branching hyphae, which produce spore-forming structures after periods of vegetative growth. They often invade niches abundant in hostile microorganisms and against which they release their antibiotics to eliminate their rivals; however, these bacteria faced a dilemma as they also were susceptible to the killers that they produced. To overcome this problem of potential suicide, these organisms had to develop resistance genes to the very antibiotics they were producing. The defence system that evolved was far less damaging than the radical mutations found in the target sites of clinical bacteria to produce clinical resistance.

Aminoglycoside-producing *Streptomyces* have evolved enzymes capable of donating onto the aminoglycoside molecules functional groups that inactivate them; these functional groups are usually phosphate, adenyl and acetate moieties and mediated by transferase enzymes. This capability to control the immediate vicinity must have evolved specifically and is mediated by so-called "secondary metabolites". These are, by definition, naturally produced substances which do not participate in the internal metabolism of the cell that produces it. Secondary metabolites, which defend antibiotic-producing bacteria, are derived from the gene clusters that are responsible for encoding antibiotic biosynthesis and regulation. Exactly how these transferase genes emerged in *Streptomyces* is unknown but their close proximity to the antibiotic production genes suggests that their evolution has been dependent on them. Recent improvements in the ability to sequence DNA have revealed that there are considerable similarities between genes within the clusters of antibiotic-producing and antibiotic-resistance genes. This suggests that the resistance genes may have arisen directly from the production genes. A likely scenario might have been that once the antibiotic production genes had evolved, the final gene in the sequence might have undergone a series of duplications. The product of this gene has an active site that can bind the antibiotic, so it may have been duplicated merely as a rapid response to increase the number of active sites and absorb aminoglycoside molecules that have not been exported. Once these duplications have occurred, they may be followed by random mutations which might give rise to a series of proteins that would certainly be able to bind the same substrate; after all, they are derived from the protein that made it, but they might also be able to donate a functional group which prevents the binding to the target and renders the molecule inactive, thus acting as a mechanism of resistance. The evolution of an enzyme that added an inactivating group to aminoglycosides and then released the product to allow it to inactivate another molecule is a much more efficient system than increasing the number of active sites to pre-occupy these dangerous molecules. Close examination of the sequences of the aminoglycoside resistance mechanisms in *Streptomyces* provides support for this theory; the genes that complete the synthesis of the amino-glycoside molecules are very similar to those that encode the transferases that inactivate the molecule.

This model could be extended further to provide resistance genes to antibiotics produced by other species and indeed many *Streptomyces* species do possess resistance genes to aminoglycosides produced by other *Streptomyces* species. These secondary metabolites are also found in other organisms, including bacteria, algae, corals, sponges, plants and some lower animals, essentially in organisms that do not have an immune system and have to rely on chemical defences. So the potential for resistance genes exists in a variety of environmental organisms, a fact that we ignored at our peril when we provided the conditions that allowed these microorganisms to invade our hospitalised patients.

The process of gene duplication and mutation could be a slow process, especially for the optimum genes to emerge. In the relatively sedentary environment of the soil bacterium, the time taken to evolve the best gene might be acceptable, but in the rapidly changing environment of the clinical bacterium this trial-and-error system with chromosomal genes would be catastrophic, since the development of appropriate chromosomal mutations may be too slow in responding to successive clinical antibiotic regimens, especially at the speed with which they are administered.

The transferase resistance mechanisms of *Streptomyces* have evolved over millions of year and are more effective than any mechanisms that have emerged in clinical bacteria during the past 50 years of the antibiotic era. Therefore, a more favourable alternative for the clinical bacterium would be to acquire these streamlined resistance mechanisms from the antibiotic producers, which could then be imported on plasmids. Comparison of sequences of the resistance genes found in *Streptomyces* spp with those encoded by plasmids in both gram-positive and gram-negative clinical bacteria reveals considerable similarities. Bacteria isolated and stored 70 years ago, before the advent of antibiotics, do not possess these plasmid resistance genes so it may be assumed that they have been imported on plasmid vectors in response to Man's release of antibiotics to cure infection.

Plasmids do not always carry genes that inactivate antibiotics, and they are certainly the easiest to identify. Tetracycline is a natural antibiotic originally isolated from *Streptomyces aureofaciens*, but this species had no transferase enzyme that could inactivate tetracycline. Indeed, on plasmids that confer transferable tetracycline resistance, there are no genes that encode transferases capable of disabling tetracycline. A different resistance mechanism has evolved; the tetracycline molecule never reaches sufficient concentration in the cytoplasm of the bacterium to inhibit protein synthesis. The chromosomal mechanism of resistance was to disable the active transport system of the cell, but a plasmid cannot achieve this; it would require more than the one gene product that is used by most plasmid-encoded resistance systems. Instead the plasmid has developed a method that pumps the tetracycline out of the cell far faster than the active transport can introduce it. The dynamics of this efflux pump is effectively to render the cell impenetrable to tetracycline. It is not a strict impermeability mechanism because tetracycline does enter the cell but to the observer that is precisely its effect. Tetracycline, bound to magnesium, crosses the outer membrane through the porin OmpF. Fast efflux is mediated by the Tet protein, which is an efflux pump encoded by plasmid-encoded *tetA* genes, and is located in the inner membrane. It works by pumping magnesium-chelated tetracycline into the periplasm. In actual fact, this resistance mechanism has the same net result as the inactivation systems used against the other antibiotics: it prevents active antibiotic binding to the target. Efflux pumps are poorly understood at the moment, though their importance is slowly being recognised. They must have evolved to remove unwanted molecules, perhaps even antibiotics, from the cell. They have usually been associated with chromosomal production; only the tetracycline resistance gene is plasmid-mediated. This is probably because the tetracycline resistance gene is the only one that has evolved adequately for efficient expression from a plasmid. Some species, notably *Pseudomonas aeruginosa*, are known to have elaborate chromosomally encoded efflux pumps, which bypass the outer membrane barrier. These pumps can excrete a series of unrelated antibiotics and thus can confer resistance to a number of antibiotics at the same time. They are known as Multiple Antibiotic Resistance genes or *mar* genes. They are a cause of great concern because, although they are not yet plasmid-mediated and therefore confined to the species in which they evolved, the potential for plasmid transmission of the genes exists. In addition, they are able to excrete and therefore cause resistance to antibiotics for which there has been, up until this time, no plasmid-encoded resistance. The potential resistance mechanism causing most apprehension is to the quinolones. It has been reported that there is a plasmid-encoded resistance mechanism to this group of antibiotics in the USA. The mechanism is unknown but chromosomal efflux pumps, notably encoding *norA*, that can actively carry norfloxacin out of the cell are known and it is possible that this is how the plasmid overcomes quinolones. The prospects for the fluorinated quinolones are now seriously jeopardised and plasmids carrying this resistance gene may herald the eventual demise of this invaluable group of drugs.

Once the source of plasmid-encoded resistance genes had been identified, it was relatively clear to understand how natural products, such as penicillins and aminoglycosides, have selected resistance mechanisms in other species. There are a number of antibacterial drugs that have no connection with natural products, so it might be far less obvious how resistance genes could be selected and then imported. Trimethoprim was engineered to inhibit bacterial dihydrofolate reductase while leaving the mammalian enzyme virtually unaffected. The modelling of this new synthetic drug was applauded not least because it cultivated the belief that resistance should be slow to develop and that plasmid-encoded resistance should be possible. The clinical introduction of trimethoprim coincided with my introduction to research science; I joined the team of an extraordinary scientist, J.T. (Johnty) Smith, at the School of Pharmacy in London to study for a PhD. Smith was interested in the resistance to sulphonamides and set me the task of examining the mechanism of plasmid-encoded sulphonamide resistance because, enigmatically, bacteria preferentially used plasmids to overcome these original synthetic antibacterials; however, 12 years after their identification the resistance mechanism had still not been found. I soon found out why; they were extremely difficult to work with and the resistance was difficult to quantify. Smith worked closely with Naomi Datta at the Royal Post-graduate Medical School, and she suddenly came across some clinical bacteria that were very resistant to trimethoprim and were able to transfer this resistance. Smith heard about this discovery and immediately rushed round to ask Datta if his "lad" could try to identify the resistance mechanism. The first I found out of this was when Smith entered the laboratory, puffing at his pipe, and told me to drop everything that I had done in the previous six months and start working on these new bacteria. Sulphonamide resistance was out, trimethoprim resistance was in.

The problem with scientific research is that the longer you work in it, the more you are the victim of your own preconceptions and prejudices. That is one of the good reasons for employing research students; they are supposed to question your dogma. Faced with the two previously discovered resistance mechanisms, I could find no evidence of either modification of trimethoprim or efflux of the drug out of the cell. Smith and I drew up a list of every possible mechanism of resistance that a bacterium or plasmid could use to confer resistance; then we started eliminating them one by one as unlikely to be viable when encoded by the plasmid vector.

The production of an additional target, less resistant to the binding of the antibiotic, was dismissed by most experts as a wasteful and impracticable mechanism for a plasmid to employ. After all, if that was the mechanism of plasmid-encoded resistance to the protein synthesis inhibitors, then the plasmid would help encode ribosomes with resistant target proteins; however, these would co-exist with ribosomes with sensitive target proteins which would continue to bind the antibiotics. The production of ribosomes and their constituent proteins constitutes an enormous proportion of the bacterium's biosynthetic production. To allow antibiotics to continue inhibiting sensitive ribosomes and the accompanying production of abortive protein molecules is far too consuming to succeed.

I had exhausted accepted mechanisms so, unhindered by convention and with the naïveté of youth, I embarked on a search to find a plasmid-encoded dihydrofolate reductase. I used to grow up to 50 litres of bacteria concentrate and then break open the cells with ultrasonics. I separated the proteins in a vain search for a new dihydrofolate reductase. While examining one of my many graphs from the dihydrofolate reductase separation on a chromatography column, Smith noticed that the peak of dihydrofolate reductase activity was not symmetrical, but had a shoulder at one side. He suggested I changed my assay technique and I found that the plasmid-containing cells produced an additional dihydrofolate reductase, albeit in minute concentrations. The additional enzyme was as capable at reducing dihydrofolate to the active tetrahydrofolate as was the bacterial chromosomal enzyme; however, it was not inhibited by trimethoprim. We purified the enzyme and found that it was quite unlike the bacterial dihydrofolate reductase and probably originated from a quite different source. However, we had found the mechanism of resistance and it was completely novel.

The mechanism of this plasmid-encoded resistance worked in a manner unfamiliar to bacteria; the chromosomally encoded target was still inhibited by trimethoprim. The plasmid enzyme simply bypassed the blockade of the chromosomal target. This mechanism had been dismissed as a possibility for conferring resistance to other antibacterials, so why should it operate for trimethoprim? There are probably two reasons. The first is that the options are probably quite limited; it would be unlikely that there are enzymes capable of modifying this synthetic compound and efflux pumps are relatively ineffective. Trimethoprim is very well absorbed when given orally and concentrates in the urine, reaching levels in excess of 100 mg per litre. If the sensitive bacterium has an MIC of trimethoprim of around 0.1–0.5 mg per litre, the bacterium has to employ a resistance mechanism that can overcome drug concentrations 1000 times that capable of killing the sensitive bacterium. An efflux pump is simply too inefficient; trimethoprim, at the concentrations in the urine, diffuses so rapidly into the cell that no pump could cope. The second reason is that trimethoprim inhibits the reduction of dihydrofolate to tetrahydrofolate, a co-factor which is the equivalent of a vitamin in human cells. This co-factor is largely a carrier of methyl groups which it can accept and donate as required; it is usually only required in quite small quantities so the number of dihydrofolate reductase molecules within the cell is low. The cell can afford to manufacture two dihydrofolate reductases if challenged with massive concentrations of trimethoprim. There is only one other antibacterial drug for which this mechanism of resistance would work and that, rather ironically, is the plasmid-resistance to the sulphonamides, the mechanism that my supervisor stopped me working on in order to study trimethoprim resistance. The mechanism of sulphonamide resistance was found a year after trimethoprim resistance and has been studied in far less detail.

In the intervening 26 years, there have been 15 plasmid-encoded dihydrofolate reductases found in gram-negative bacteria. Many of them confer extremely high levels of resistance, raising the capability of the bacterium to resist trimethoprim at concentrations 10,000 times greater than that sufficient to inhibit the sensitive plasmid-free cell, more than enough to deal with trimethoprim concentrations in the urine. Plasmid-encoded trimethoprim resistance in pathogens such as *Salmonella* and *Shigella*, which invade the gut, is often at a much lower level. It may be with MICs around 50 mg per litre, well capable of dealing with trimethoprim concentrations in the intestine. The source of these new dihydrofolate reductases has been the cause of some speculation; there has been no soil mileu to select out the resistance genes over aeons. However, dihydrofolate reductase genes are universally distributed in almost all cell types, human, plant and bacterial. They are even carried by certain viruses, so the choice of potential genes for the plasmid has been wide. So it would not be too difficult to envisage a scenario where an insusceptible dihydrofolate reductase gene in a distant species was mobilised and transported to clinical bacteria undergoing trimethoprim challenge; however, plasmid-encoded trimethoprim resistance emerged only two years after the clinical introduction of the drug, and extensive mobilisation of dihydrofolate reductase genes, from evolutionary distant sources, seems unlikely. Examination of the plasmid-encoded resistance genes suggests that the origin had been bacterial and that they might have been derived from the very species that they now found themselves.

Evidence for this came from an unlikely source. A recent PhD graduate from my laboratory, Hilary-Kay Young, went to India to investigate massively high incidences of resistance in common pathogens. She returned to Scotland with a suitcase full of well-categorised clinical bacteria resistant to trimethoprim. While she was painstakingly investigating these isolates, she noticed that some conferred only low levels of resistance; however, the resistance was transferable and plasmid-mediated. She then made a crucial observation: the level of trimethoprim resistance was dependent on the concentration of drug to which the bacteria had previously been exposed. When she measured the level of dihydrofolate reductase within the cell, she noticed that it increased to 600 times its level in unchallenged cells. The dihydrofolate reductase gene was capable of induction, an extremely rare condition in plasmid-encoded genes. She determined the biochemical properties of the dihydrofolate reductase and found that it was only marginally less sensitive to trimethoprim than the *E. coli* chromosomal enzyme. The mechanism of resistance, on trimethoprim challenge, was by a rapid increase in the number of dihydrofolate reductase molecules, and hence active sites to soak up the antibacterial drug, as well as a bypass by an enzyme that was slightly less sensitive. We appeared to be witnessing the actual development of a plasmid-encoded trimethoprim resistance mechanism; we were at the stage where most of the resistance was mediated by a marked increase in active sites to cope with the immediate challenge, rather than a major alteration of the active site. This seems to be the first stage in some resistance development.

When Hilary-Kay Young examined the DNA sequence of this dihydrofolate reductase, known as the type IV, she found that it was similar to the *E. coli* chromosomal enzyme and probably had originated from it. She may have been observing one of the blind alleys of evolution – the gene had become caught up in a cluster of induction genes – but what she had demonstrated was that the chromosomal gene can migrate to the plasmid and undergo crucial mutations to desensitise it to the binding of trimethoprim. It is still not clear whether the high-level plasmid-encoded trimethoprim-resistant dihydrofolate reductase are derived directly from the type IV enzyme but they do seem to have undergone some similar series of mutations. The gene would have to duplicate and increase its expression to increase the number of active sites as an early response to trimethoprim challenge. Thus follows the now familiar story of duplicated genes undergoing further mutations to increase the resistance level, in this case, by further decreasing the ability to bind trimethoprim but retaining the ability to reduce dihydrofolate.

Most of the plasmid-encoded genes have now been sequenced and they largely fall into two groups, those possessing a sequence similar to the type Ia and those related to the type II. The type Ia enzyme is one of the most common plasmid-encoded dihydrofolate reductases and has a very distinctive set of biochemical properties, which seem particularly successful in conferring trimethoprim resistance. Other successful plasmid-encoded dihydrofolate reductases in gram-negative bacteria have similar biochemical properties (i.e. type V, type VII, etc.), suggesting that a particular compendium of activities is required to promote the most efficient resistance.

Plasmid-encoded trimethoprim resistance in gram-positive bacteria was much slower to develop, more than 10 years after it appeared in gram-negative bacteria. As the so-called methicillin-resistant *Staphylococcus aureus* spread from hospital to hospital, particularly in Australia and the United States, its plasmids acquired resistance genes to trimethoprim. Like their gram-negative counterparts, these genes conferred very high levels of resistance so that the host bacterium could survive in 1000 mg trimethoprim per litre. Despite the wide dissemination of gram-positive bacteria throughout the world, there is only one plasmid-encoded dihydrofolate reductase, the type S1 in staphylococci, which is found in virtually all species. This enzyme is very similar in size to the type Ia in gram-negative bacteria and, for the most part, it has very similar properties so it might be thought to have the same origin; however, the source of this plasmid-encoded dihydrofolate reductase gene was much more obvious than the gram-negative enzymes. The type S1 dihydrofolate reductase gene has been sequenced and has been shown to be very similar to the chromosomal dihydrofolate reductases of gram-positive bacteria, particularly to that of *Staphylococcus epidermidis*; it differs by four nucleotide changes in the gene which translate to alterations at crucial binding positions for trimethoprim. Here is a clear example showing the role of a chromosomal gene as the progenitor of a plasmid gene. The surprising fact could be that only four changes are required, so why then did it take so long for trimethoprim resistance to become plasmid-encoded and why is there only one type? Closer examination of the gene would show four distinct mutations working *in concert*; none of the mutations give much of a selective advantage on their own. So when trimethoprim first challenged *S. epidermidis,* the dihydrofolate reductase gene presumably duplicated to respond to this threat. It might then have mutated but the resultant mutant did not confer a sufficient advantage. When the much rarer event of two mutations occurred simultaneously or one soon after another, sufficient advantage might be achieved; however, under normal events this would be expected to be very infrequent. If the double mutation conferred some advantage then it might persist and further mutations occur after continual challenge. The chances of double mutations are very remote, perhaps odds of 1 in 100,000,000,000,000,000. In the normal event, this would take approximately 1 million infections, each producing 1 billion bacteria, before this double mutant emerged. The dynamics of the delay in the emergence of quite simple mutational changes suggest that two mutations were required before any advantage was achieved. This was different from the extended-spectrum β-lactamases where each mutation conferred an advantage, so the potential for the emergence of favourable mutations occurred at each infection treated with a cephalosporin.

The gram-negative dihydrofolate reductase differed from the gram-negative enzyme in that it did not confer resistance to methotrexate. Trimethoprim is a structural analogue of the substrate of the enzyme, dihydrofolate; actually it is not all that similar, it is considerably smaller, but the bacterial chromosomal enzymes recognises it as similar, though the mammalian enzyme does not; hence it is selective. Methotrexate is a general dihydrofolate reductase inhibitor; it has no preference for the bacterial enzyme and binds to mammalian dihydrofolate reductases just as readily. It is a much closer structural analogue of dihydrofolate than trimethoprim. Methotrexate binds to the gram-positive dihydrofolate reductase as it would to any dihydrofolate molecule; however, it does not bind the active site of many of the gram-negative plasmid-encoded dihydrofolate reductases. This suggests that these enzymes are more evolved and more sophisticated than their gram-positive counterparts. It has a much greater implication because it may be predicting that it would be impossible to devise analogues of trimethoprim that can overcome the plasmid-encoded resistance of gram-negative bacteria. Methotrexate is the least selective of all dihydrofolate reductase inhibitors because it is structurally so close to dihydrofolate; however, the plasmid dihydrofolate reductases can distinguish the two for they allow dihydrofolate to bind at rates similar to the chromosomal enzyme but they exclude methotrexate. Therefore a new analogue is likely to have to be closer to the structure of dihydrofolate than methotrexate if it is to bind to the active site, but if it is closer in structure than methotrexate, what chance has it to be selective against the bacterial chromosomal enzyme? This is a real catch-22 problem. The inability of the pharmaceutical industry to produce a single trimethoprim analogue that can overcome the plasmid-encoded enzymes suggests that this puzzle is truly unsolvable.

TRANSPOSONS

A nagging question is what is the most consequential element that determines the diversification and spread of resistance genes? I was originally trained as a biochemist so I often favour the view that it is the efficiency of the biochemical properties of the gene product that determines success, though a molecular biologist might suggest that the genetic carrier, on which the resistance gene is located, is the critical factor. Plasmid carriage of resistance genes does explain why some resistance genes are more prevalent than others; simply they move between bacteria. Direct plasmid carriage was considered to be the sole mechanism for transmission of mobile resistance genes for 15 years; however, by the middle of the 1970s, this view of gene movement was insufficient for the observations made. Plasmids did not emerge because of the use of antibiotics and the development of resistance genes. At that time, a detailed examination was being carried out on a clinical bacterial population from Canada that had been stored towards the end of the First World War, at least 10 years before the antibiotic era. If antibiotics had selected plasmids, these bacteria should have none; however, there were as many bacteria carrying plasmids as there now are in current clinical bacteria and most of these plasmids were closely related to those that we currently find. The only difference between modern plasmids and those present 75 years ago is that the latter carried very few resistance genes. Plasmids are the natural partners of clinical bacteria but how did they acquire resistance genes? Pressing evidence suggesting some form of gene exchange between plasmids came from observations of the TEM-1 β-lactamase. Soon after its initial emergence, this enzyme was suddenly found in a myriad of different bacteria and the dynamics of its spread were difficult to explain by plasmid dissemination alone. Geneticists were finding the same gene in plasmids of different incompatibility groups, but how was it moving from one plasmid to a completely unrelated type? The answer came from the emerging science of molecular biology: the genes appear to transpose themselves from one plasmid to another. When molecular biologists measured the size of the plasmid DNA before and after acquisition of the resistance gene, they could measure a physical increase in size. Plasmids were exchanging and sharing the TEM-1 β-lactamase gene and they also carried the genes that allowed this recombination.

The plasmids of clinical bacteria appear always to have had mobile segments of DNA that could duplicate and insert copies into other plasmids. The mobile genetic segments are known as insertion sequences, though it not clear what their original role had been. If two insertion sequences locate in reasonably close proximity to one another, along the same DNA strand, then they retain their ability to relocate. They do this as a complete unit so that the DNA caught between the two insertion sequences is also mobilised. This might have been the original role of insertion sequences to move genes from one plasmid to another but now, of course, they can enclose resistance genes. This is known as a transposon and immediately offered an explanation for the apparent promiscuity of certain resistance genes; they were imprisoned between very proficient insertion sequences. So the plasmids from 75 years ago, which contained few, if any, resistance genes, acquired them by sequential contact with DNA molecules containing transposons, which were able to move a copy across to the new plasmid. Transposons and transposition may give an explanation, at least in part, as to how the resistance genes moved from the original antibiotic-producing bacteria to plasmids in clinical bacteria. It certainly explains the ubiquity of some genes and their presence in many different plasmid types. It is quite easy to see how this might work in practice. A transposon encoding a resistance gene may be located in a plasmid specific to *Vibrio cholerae*. In this condition, the transposon would be constrained by the limitation of its host plasmid; it would always be in *V. cholerae*. At some time, the *V. cholerae* strain comes into close contact with an *E. coli* carrying a plasmid with a much broader host range. The transposon moves onto the broad-host-range plasmid, which can migrate into other species; in one leap our plasmid becomes free of *V. cholerae*.

Transposition not only allows the rapid dissemination of resistance genes between plasmids but it also increases the adaptability to respond to changes in environmental conditions. If, for example, a strain carrying a plasmid was suddenly starved of nutrients, then the carriage of the extra plasmid DNA may be a burden that places the bacterium at a competitive disadvantage. It is unlikely that the plasmid will physically be expelled from the cell but rather plasmid-free cells will have a survival advantage in these harsh conditions. Loss of the plasmid would mean loss of the transposon. Most transposons can move between any replicon, so they can migrate between plasmid and the bacterial chromosome. So a transposon that has previously migrated, and inserted a copy of itself into the bacterial chromosome, will survive when conditions become austere. Some might argue that the austerity might be the trigger that stimulates the transposition into the chromosome; there is no direct evidence for this but certainly a transposon that had the capacity to jump to the chromosome at the first sign of trouble would be at an enormous selective advantage. Once in the chromosome, the transposon can survive virtually indefinitely, replicating every time the chromosome replicates. The host bacterium benefits because it now has the mechanism to survive in the presence of certain antibiotics, permanently embedded into its own genetic material. As the environment changes and the availability of nutrients improves, the cell is likely to be visited by plasmids again. The transposon can transpose itself from the chromosome onto the new plasmid and then move out of the cell to new species.

The capability to transpose DNA between plasmids and the chromosome is clearly important. A number of completely different mechanisms have evolved to achieve this ability. There are two major classes of transposon, class I and class II. Class I transposons are often referred to as composite or compound transposons. They are composed of a middle region of DNA containing genes that may of be use to the host, often antibiotic resistance genes. This central region is bordered by the two insertion sequences. Associated with each of these insertion sequence elements is the gene to encode an enzyme, called a transposase, but only the transposase from one of these elements is ever expressed; the other gene is disabled. The transposase provides the capability to move for the whole transposon. This type of transposon appears to migrate by a non-replicative mechanism which extracts the whole element out of the DNA and then reinserts it into the new replicon. The transposition event is strictly timed to occur immediately after replication of the transposon during normal DNA replication of the donor replicon. This linking of transposition to follow replication ensures that the loss of the donor DNA does not compromise the survival of the transposon, the host plasmid or host bacterium.

Class II transposons or complex transposons behave in a completely different manner. The resistance genes in these transposons are located at one end of the transposon element and the transposition event is mediated by two enzymes, not one, and a short DNA sequence, known as the *res* site, which acts as a focal point to complete the transposition process. The first enzyme is a transposase but it works in a quite different manner from the transposase in class I transposons. This enzyme forms a cointegrate intermediate and replication of the transposon occurs during cointegrate formation. The donor DNA element and recipient DNA element are physically bridged by the transposon, they are linked together by the two single-stranded DNA strands of the transposon. The second enzyme, encoded by the transposon, is a resolvase. The role of this enzyme is to separate the cointegrate into its component parts. This it achieves by completing the DNA synthesis along the single strands of the transposon and then, by site-specific recombination between the *res* sites on the two transposons bridging the cointegrate, the individual replicons are released each with a copy of the transposon embedded in them. The ubiquitous TEM-1 β-lactamase is invariably found on a class II transposon, Tn*3*, but another particularly interesting class II transposon is Tn*21*. This sub-class of transposons is widely distributed in clinical bacteria around the world; however, it possesses even greater flexibility in the ability to disseminate resistance genes, as it can also carry an integron.

There is a so-called class III transposon but these really are not a single class at all. They are a number of quite unrelated transposons, with often unrelated transposition mechanisms, which are currently poorly understood. Some of them have a myriad of transposition genes; the functions of most are unknown and may not even be essential.

The fourth group of transposons are completely different and may share the name transposon merely because they are unable to replicate on their own. They are called the conjugative transposons. In fact, they may be much more similar to plasmids than transposons. Conjugative transposons are usually larger than non-conjugative, ranging in size from around 20 kb to 150 kb, very similar in size to many plasmids. Like plasmids, they have the capability to exist as covalently closed circles of DNA, independent of the chromosome or true plasmids. Unlike plasmids, however, they have no means of replication so there are no maintenance genes and no genes to restrict copy number within the cell. So conjugative transposons can be present in many copies within the cell. Conjugative transposons can migrate from one DNA molecule to another within the cell, like a true transposon, but this is by a process quite unlike any other transposition event. In fact it is more akin to the integration of temperate bacterial viruses (bacteriophages) into bacterial chromosomal DNA. These "transposons" do not have their own recombination system so are much more reliant on the bacterial functions than the true transposon. Like plasmids, these transposons can migrate directly from one cell to another by conjugation; they have all the necessary genes and can replicate the DNA during the conjugation event. This capability significantly increases the flexibility to spread; these elements are capable not only of integrating into other replicons but of migrating between replicons as well, and, being able to transfer themselves between cells, should have the greatest flexibility of all. It is difficult to establish exactly what they are, whether they are plasmid that have lost the capability to replicate or some transposon-like element that has acquired the genes for transfer. I suspect that the former is more likely. The carriage of conjugation genes is expensive and may be the reason why these elements are found comparatively rarely. At first, they were thought to be confined to gram-positive bacteria but they have now been found in a few gram-negative species and it is likely that they may be present in most clinical bacteria.

The success of true non-conjugative transposons comes from the success of the insertion sequences. Several factors determine the ability of these elements to mobilise resistance genes; the first is the insertion specificity of the element. Some transposons, and thus their insertion sequences, are highly specific. They only insert within defined nucleotide sequences but transposons with such limited sites for transposition are unlikely to be significant in the formation of compound transposons. On the other hand, insertion sequence elements that transpose randomly and can enter at any position within a nucleotide sequence are also probably likely to be unsuccessful because insertion could occur as easily within a resistance gene as it would in the flanking regions. Some insertion sequence elements, such as IS1, regularly flank compound transposons and they have been found to enter DNA sequences that have an abnormally high adenine or thymine content. The regions of DNA most commonly rich in these nucleotides are the promoter sections, which are located next to the gene but never within it. Therefore an insertion sequence that has a preference for these regions would always insert outside the gene. These insertion sequence elements are more likely to form the ends of compound transposons than insertion sequences with random insertion sites because they would be less likely to interrupt the transcription of the resistance gene and thus mobilisation of a functional gene may be more successful.

After one insertion sequence element enters the promoter region upstream of the gene, a second IS element must insert downstream of the gene so that the gene may be mobilised. Again there appears to be some direction for this; acquisition of one insertion sequence element increases the probability that a second insertion sequence element will insert at a nearby site. This is thought to be the mechanism for the insertion of IS1 and its apparent success. This may not be the case with other insertion sequence elements; the presence of any DNA sequence automatically attracts other closely related DNA sequences, especially if they are carried on a second plasmid. Once these are in close proximity, transfer may occur by homologous recombination between a plasmid carrying an IS element adjacent to a resistance gene and the second insertion sequence element. This recombination would result in the formation of a cointegrate harbouring two copies of the insertion sequence element flanking the resistance gene and a large region of intervening DNA. The intervening DNA is likely to be lost by deletion which would increase the probability of the formation of a stable compound transposon.

This hypothesis may illustrate how two insertion sequence elements can flank a resistance gene to form a transposon but it does not account for the extreme promiscuity of these elements as they disperse rapidly through the plasmid population. A paradox arises; the insertion sequence is a promiscuous element in its own right, so why should it lose this inclination when two of them form a transposon by flanking a resistance gene? There has to be some mechanism that can suppress the mobility of the individual insertion sequence element and shift this talent to the whole transposon. Studies on the two compound transposons known as Tn5 were performed to reveal the switch to co-ordinated transposition of the entire compound transposon rather than independent transposition of the insertion sequence flanking elements. The internal ends of the insertion sequence elements in Tn5 have been modified. These regions normally serve as focal points for transposase activity but they have been mutated and are no longer recognised as substrates for the transposase enzyme. In addition, some of the nucleotides in the internal ends of these insertion sequence elements have been modified; methyl groups are added to some of the nucleotides by a Dam methylase. Methylated nucleotides are not recognised as sites of transposition. Modification of the internal repeat sequences by both mutation and modification radically increases the possibility that the newly formed compound transposon is transposed as a whole unit. The transposase gene in one of the insertion sequences is mutated to disable it. In the case of Tn5, this transposase is traditionally depicted on the left of the transposon.

This model provides a reasonable clear model of how resistance genes are captured between insertion sequences to become part of compound transposons. It does not, however, explain how transposons procure resistance genes present in the chromosome of inherently resistant species and provide the means for them to gain wider access to clinical bacteria and their plasmids. Analysis of the DNA nucleotide sequence of some of the resistance gene carried within compound transposons suggests that they are derived from the chromosomal genes of clinical bacteria. The Tn5 encodes an aminoglycoside phosphotransferase, conferring kanamycin resistance, which is related to chromosomally encoded kanamycin resistance genes found in many gram-negative and gram-positive pathogens. We have already seen the similarity of the plasmid-encoded dihydrofolate reductase genes encoding trimethoprim resistance to those in the bacterial chromosome. Most of the plasmid-encoded genes reside within transposons. Although not yet assigned to the class C β-lactamase, BIL-1, is likely to have been extracted out of the chromosome by the same method.

Insertion sequence elements are widespread in most clinical bacteria, both gram-negative and gram-positive. It is argued that the antibiotic-producing bacteria are the source of many plasmid, and thus transposon-encoded, resistance genes but *Streptomyces* spp, by far the most prolific of these producers, is not a favoured host of these elements. Indeed, in some species, none have ever been found. This begs the query, if they are required to extract resistance genes out of the chromosome, how is this achieved in *Streptomyces* spp?

INTEGRONS

The search for mobile and mobilising elements was not over. As more sophisticated techniques became available to study resistance genes, certain anomalies were found. Comparison of the DNA nucleotide sequences of Tn*21* transposons revealed that there were regions of absolute conservation which flanked variable DNA regions. The variable regions encoded different resistance genes. A new type of genetic element had been identified; it is called an integron and is characterised by conserved 5' and 3' ends which flank a variable central DNA segment. Most characteristically, the 5' conserved end contains a functional gene, called *intI*, which encodes for an enzyme, labelled an integrase, that gives the whole element its name.

Overall, the integron appears to possess the ability to poach resistance genes from other DNA molecules. This might be other plasmids or even the bacterial chromosome. It is the role of the integrase to mediate the insertion of the "foreign" DNA sequences, which are known as gene cassettes, into a specific attachment site called *attI*. This inserted DNA becomes the variable region of the integron. At the other side of the variable region from the integrase gene, in the 3' conserved region, there are two genes whose function is still unknown and a gene encoding sulphonamide resistance. The gene cassette inserted into the central variable region comprises a complete gene, often encoding antibiotic resistance, which is flanked by a 7-base conserved sequence upstream of the gene and a sequence of nucleotides known as the 59-base element located 3' to the structural gene. Once one cassette has been inserted at the *attI* site, the integron can search for other gene cassettes and can insert these next to the first so the integron can build up a series of gene cassettes within itself. There must be some limit on the number that can be included though this has not yet been found.

The gene cassette in the variable region of the integron can also be excised by the integrase and they can be found as circular molecules within the cytoplasm of cells. In this state, they are not able to replicate and must be reinserted back into an integron structure to ensure maintenance and survival. The integrase is responsible for the process of integration of "foreign DNA" but the ability to integrate into integrons, and to be excised from them, relies on the presence of the 59-base element. This element, which paradoxically is often not exactly 59 bases in length, always has sequences that represent inverted moderately homologous repeat sequences. These vary in length from one cassette to another but all have the potential to form stem-loop structures. Despite the variability of the sequence of the 59-base elements, they all possess a highly conserved 7-base sequence (GTTRRRY) which is the label for the insertion site. The site-specific recombination of the integrated sequences occurs within or on either side of the GTT sequences found both within the cassette that is integrated and the 59-base element resident within the integron. In order to become integrated the gene cassette has to possess a 3' 59-base element which lacks both the last 7 bases and a conserved 7-base sequence 5' to the structural gene.

Integrons are newly discovered harbingers of antibiotic resistance and their origins can only be speculated. It is thought that the early integrons, from which the current integrons are derived, comprised only the 5' conserved segment and a reduced 3' segment.

The only requirement for resistance genes to become inserted within an integron is to be closely attached to a 59-base element. The distribution and occurrence of 59-base elements within clinical bacteria are not yet known but preliminary studies indicate that such elements are very common in most species of enterobacteria so it is possible that many of the resistance genes commonly found in clinical bacteria may have become mobilised from their original host after connection of the gene with a 59-base element. Indeed many of the resistance genes found within integrons are either significantly homologous or even identical to chromosomal genes in *Enterobacteriaceae* and *Streptomyces*. Currently, there has been insufficient sequencing of bacterial chromosome DNA; however, when this has been achieved, we may see that these resistance genes are also associated with 59-base elements and whether these gene cassettes have evolved because antibiotic selection pressure has forced the mobilisation of resistance genes to other bacterial genera.

Integrons may carry more than one genetic cassette and a myriad of many different resistance genes have been found within integrons. The most common are the genes coding for aminoglycoside modifying enzymes, some of the rarer β-lactamase genes (notably OXA and PSE), trimethoprim resistance genes (encoding the types Ia, V, VII, X dihydrofolate reductase), and chloramphenicol resistance genes (the cml efflux pump). It is important to note that certain, highly successful resistance genes, such as the TEM β-lactamase genes and the aminoglycoside phosphotransferase genes, have not so far been found within integron structures. This is possibly because they are already located within highly successful resistance transposons which ensure their spread and survival within a large variety of bacterial species.

The whole concept of integrons begs the question as to whether the acquisition of resistance genes is random and if Darwinian selection ensures that only those that are useful are selected. Even though there may be a massive number of potential integrons within clinical bacteria, it seems that a random selection procedure would just be too inefficient. Does the integron acquire resistance genes in response to specific challenges? We have only just been able to measure the physical act of integration so it is too early to tell, but "directed" evolution seems likely.

In the past 35 years, we have witnessed the whole era of mobile resistance genes. Plasmids were identified in the late 1950s, transposons in the mid-1970s and integrons in early 1990s. The genetic carriers were identified as techniques improved; integrons could never have been discovered earlier, which questions what other carriers might already be there but we simply do not have the expertise or understanding to identify them. It is likely that we shall soon witness how resistance development is directed by the challenges imposed upon bacteria but before we do this we need to know the conditions that promote the emergence and spread of resistance. We all know that resistance is a problem but we still only barely understand why resistance emerges in the first place. An understanding of the epidemiology of resistance emergence and spread would answer this.

6

Identifying the enemy – the safety of antibiotics

We all may have sat apprehensively, at one time or another, in the waiting-room of the doctor's surgery waiting to find out whether the sore throat, chest infection, festering wound, carbuncle or simply burning while passing urine is the sign of some hidden menace that is going to develop into something much more serious. How relieved we become when we find that all that is required is a simple course of tablets. The pressure on the general practitioner to prescribe some pharmaceutical is often great, particularly if the patient is a child. If the patient is left untreated and deteriorates, the physician could be considered negligent. On the other hand, if the patient was given a prescription, the physician had at least tried to deal with the problem. The choice of therapy might be inappropriate and ineffective but, at least, an attempt appeared to have been made to effect a cure.

Antibiotics are considered to be safe and can thus be given freely without reference to their side-effects; however, no antibiotics are safe and some can cause quite significant and even fatal side-effects. During the discovery of the early antibiotics, the problem that faced those keen to try these new compounds in the treatment of human infections was toxicity. In those pioneering days, the choice was fairly simple; the patient was likely to die unless treated so there was little to lose, though some did die from the effects of therapy rather than the infection that it was designed to treat. It was considered that some quantification of the toxicity should be established so that the true selectivity of these drugs could be determined.

In 1911, Paul Ehrlich tried to use arsphenamine to treat syphilis but found that it was not selective enough to be given in large doses; he had to give it in a series of smaller doses to limit the damage to the patient. Ehrlich tried to quantify the relationship between the increased affinity for the pathogen and the toxic level of the drug. He devised the *Chemotherapeutic index* which he defined as:

$$\frac{\text{Concentration of drug giving the minimum curative dose}}{\text{Concentration of drug giving the maximum tolerated dose}}$$

Ehrlich showed that a compound that cured trypanosomiasis in mice at a concentration of 2 mg per kg but did not kill below 50 mg per kg would have a chemotherapeutic index of 2/50 or 1/25. This is a true measure of selective toxicity.

The difficulty with determining this index is the measure of toxicity; the ultimate effect of any drug is lethality. The difficulty with taking any population of animals and measuring the concentration of drug that will kill all of them is that a few individuals are very much more sensitive to the effects of the drug and a few are very much more resistant. The lethal response of an animal population to a drug usually follows a Gaussian distribution. The variation amongst individuals is least when the median of the population is considered. So the concentration of drug required to kill 50% of the individuals varies far less than the concentration needed to kill the whole population. Therefore the LD_{50} (lethal dose required to kill 50% of the test animals) became the standard measurement. Similarly the dose of drug required to cure all infections in a population can vary wildly because of the variation between individuals; it only needs one individual to respond poorly to treatment and all the results are skewed. Therefore the CD_{50} (curative dose require to cure 50% of test animals) was devised. These measurements were substituted into Ehrlich's equation but, as Ehrlich devised it, it is a rather cumbersome method of expressing relative drug tolerance so the reciprocal is taken:

$$\frac{LD_{50}}{CD_{50}}$$

This equation provides a multiple rather than a fraction and is much easier to comprehend. It is now commonly referred to as the *therapeutic index*.

The testing of dosage levels for lethality will appear, to some, as a very Draconian technique for measuring antibiotic safety. In this era of much greater awareness of animal welfare and rights, the sacrifice of significant numbers of animals will seem immoral and indecent. One of the reasons I became a microbiologist was because I found the experimentation on animals unpleasant. Unfortunately, no suitable alternative to this type of toxicity study has been found. There are a number of tests that can determine whether a new drug is likely to be a carcinogen (cause cancer) or a teratogen (cause birth defects) which do not involve the use of animals and ironically use bacteria instead; however, the ultimate test of lethality cannot be simulated and a living mammal is required. Although this chapter will reveal some problems with certain antibiotics, it should be remembered that, in general, they are safe and this assurance of safety has been achieved because they were tested rigorously in animals before humans.

The therapeutic index gives the developers of chemotherapeutic drugs a quantifiable measure of safety; the larger the figure, the safer the drug. The early drugs of Ehrlich gave therapeutic indices of little better than 1 but once the sulphonamides and then the true antibiotics were tested, the concentrations required for LD_{50} were so high that it became impossible in many cases to measure. The best sulphonamides appeared to be tolerated in large doses and when penicillin was introduced, the drug seemed to be able to be given in unlimited concentrations without any lethal effects. As will be seen later, both drugs can be lethal under some conditions but not under the protocol of this test. As drug developers became more proficient at producing safer drugs they have developed more rigorous tests and compare the LD_1 with the CD_{99}, the concentration needed to kill 1% of the test animals against the concentration needed to cure 99%. This did, to some extent, alleviate the necessity for employing large animal studies.

The therapeutic index gives an indication of what properties need to be exploited during the development of an antibiotic. There was no point in finding a drug with twice the toxicity unless the curative dose was more than halved; the two had to go in hand-in-hand. It has been estimated that more than 5000 natural products from bacteria have now been identified as having antibacterial properties but no more than 50 of them have ever been exploited for clinical use; the rest have faltered merely because their therapeutic indices were insufficient large. It is difficult to put an absolute figure on what an acceptable therapeutic index might be; it depends on the severity of infection. All antibiotic therapy *should* balance the benefits of treatment against the injury to the patient. In severe infection this equation might be an easy one to balance and antibiotics with a therapeutic index of 4–8 are still being used. In more common and non-life-threatening situations the therapeutic index must be very much higher. Although it sometimes cannot be measured because the lethal dose is not easy to establish, it should be remembered that all chemicals may ultimately be detrimental. Their effects may not be obvious during a single challenge but may manifest themselves during either continuous or repeated therapy. Observance of this should ensure that antibiotics are never prescribed when they are not truly indicated and certainly not given to cover infections that clearly result from a virus infection. These two mistaken indications would include almost all gastrointestinal infections and most sore throats, treatment areas where hundreds of kilograms are prescribed annually.

Sulphonamides and especially penicillin had comparatively good therapeutic indices but these were only active against infections caused by gram-positive bacteria. Both drugs inhibited a biochemical step in bacteria that did not exist in mammalian cells and thus had the greatest selective potential. There was no conceivable reason why the mammalian cells should be attacked as they did not seem to possess a suitable target. Finding drugs that were as selective proved much more problematic for gram-negative bacteria; the first antibiotic for general use was streptomycin. Although the actual target of this antibiotic is still under question even 50 years after its discovery, most opinion favours the primary action of the drug to be attachment to the S12 protein on the 30S subunit of the bacterial 70S ribosome and thus it initiates abortive protein synthesis in the bacterial cell. (The S value is an arbitrary measure of size.) Mammalian cells also synthesise proteins and the machinery that they have evolved is very similar to bacteria; they use larger 80S ribosomes which are composed 40S and 60S subunits. The fact that the selective toxicity of streptomycin comes from preferential binding to the bacterial target rather than the mammalian, in contrast to the lack of a mammalian target, suggests that toxicity would be a greater problem. Toxicity is a problem for streptomycin and thus it was discontinued when safer anti-gram-negative drugs, with higher therapeutic indices, were developed. It has, however, retained a role in the treatment of tuberculosis. The aminoglycosides have been one of the pivotal groups of antibiotics and gentamicin, tobramycin and amikacin are still extensively used to treat serious hospital-acquired infection. With therapeutic indices hardly into double figures, gentamicin, in particular, has to be used with great caution. Its main adverse effect is otoxicity; it damages the eighth cranial nerve which causes hearing loss. This may be caused by single high doses but is much more likely to result from sustained treatment at lower doses. This can be either from prolonged treatment or from a series of treatment courses. The adverse effects are thus cumulative but they can also be exacerbated by concurrent use of diuretics, often required in some seriously ill patients. Unfortunately gentamicin can also cause damage to the kidneys; this nephrotoxicity can be both mild and severe dependent on dosage levels but, unlike ototoxicity, is often reversible once treatment is stopped. With so many toxicity problems, it might be difficult to justify gentamicin usage. The problems is that there are some hospital infections that respond well to this drug only. The problems have been alleviated slightly by the introduction of tobramycin but some consider that this may not be as effective in clearing bacterial infection.

These aminoglycosides cause a medical dilemma; they are clinically invaluable and irreplaceable but unless used cautiously they can be pernicious. These drugs rapidly become perilous if they are not excreted normally through the kidneys. During a course of treatment, after a single dose of drug is first given, the concentration increases in the serum (blood) until a peak is reached, called the concentration maximum or C_{max}. At this point the excretion of the drug is faster than it can build up in the serum so the concentration begins to fall. In fact the decrease is proportional to time, the time taken to reduce the concentration by 50% is constant and is known as the half-life or $t_{1/2}$. In normal courses of treatment, when it is estimated that the decaying concentration approaches the MIC, a second dose is given. At this point the concentration reaches a trough before it begins to climb again boosted by the second dose. So the cycle continues between peaks, troughs and redosing. It is possible, at the regular dosing intervals used in hospitals, to predict when the peaks and troughs should occur.

Often seriously ill patients may have kidney damage, the antibiotic therapy may actually have caused it, but the result is that the drug is not excreted normally through the kidneys. Thus when the peak is reached after the first dose, the concentration does not decay. The second dose of antibiotic further increases the concentration to form a new and much higher peak. With a potentially toxic antibiotic such as gentamicin, it would not require many antibiotic doses, under these conditions, to cause serious damage. Therefore, with drugs such as these, the concentration of antibiotic in the serum is monitored. When antibiotic concentrations were first assayed, this was usually done by a laborious bioassay that measured the drug activity by its ability to inhibit a test bacterium. Measurements were made to determine both the peak and trough serum concentrations; however, it is recognised now that sufficient information about the clearance of the drug can be obtained by measuring just the trough concentrations. The necessity to perform these measurements still exists to this day but they are now performed automatically in machines that measure the concentration of drug biochemically. The toxicity of gentamicin and the other aminoglycosides is often considered the extremes of toxicity though very recent developments to remove impurities from drugs like tobamycin appear to improve the toxicity profile markedly.

The safest group of antibiotics has always been considered to be the β-lactams, penicillin and cephalosporins. Modern versions of these antibiotics tend to be oral and have been made robust enough to pass through stomach acid. They all derive from antibiotics that originally were not acid-stable and were poorly absorbed through the gastrointestinal tract. The overcoming of the former did little to promote the latter. Thus absorption of ampicillin and amoxycillin, for example, is incomplete and probably, at best, 55%. This would mean that nearly half of the administered drug remains in the gut and is eventually excreted by this route. Before excretion, the antibiotic has to pass through the lower bowel. The lower bowel or colon is where the majority of the normal bacteria, both aerobic and anaerobic, of the gut reside. The presence of the broad-spectrum antibiotic causes widespread destruction of the aerobic bacteria. The remaining anaerobic bacteria may cause diarrhoea and this found in about 20% of treated patients. Usually it does not comprise more than loose stools; however, it can be exacerbated if the penicillin is given with a β-lactamase inhibitor. In severe cases the total destruction of aerobic bacteria may lead to pseudomembranous colititis. The removal of competition by these bacteria allows overgrowth of the anaerobe *Clostridium difficile*; the toxins that it produces cause necrotic lesions in the colon mucosa. There is also a build-up of fibrin deposits and necrotic material as well as an acute inflammatory response. This produces large pseudomembranes, hence the name, attached to the colon mucosa. The patient feels acute abdominal pain and has severe diarrhoea and in some acute cases, there may also be rectal bleeding. Removal of antibiotic treatment is often sufficient and the symptoms disappear within a few days, though failure to act on these symptoms can result in toxic dilation, perforation of the gut wall and peritonitis. The patient may be treated with antibiotics that are specific for the anaerobe, usually metronidazole or vancomycin, but anti-diarrhoeal drugs are detrimental and problems are caused if the patient starts administering them before seeking medical attention.

Penicillin treatment is sometimes associated with rash formation and may affect 5% of patients. The most severe of all reactions is hypersensitivity. Most patients know if they are allergic to penicillin and it is a question often asked before penicillins are prescribed. Some vulnerable patients always carry a bracelet or a locket which states the risk that they run if given penicillin. Its exact cause is not well defined but has often been attributed to some impurities in the production process of the antibiotic that remain in the final product. The principle of the hypersensitivity is that the patient is treated initially with a penicillin, which produces Immunoglobulin E (IgE) antibodies. Antibiotics are small molecules, usually below the size threshold that normally triggers an immune response, which was the reason for supposing that it had to be an impurity. When the patient is treated a second time, the dose of antibiotic appears to be a massive challenge to which the IgE respond. The patient is merely a host in this enormous shift of antibodies binding to the incoming drug and can go into anaphylactic shock. This may be manifested by hypotension (lowering of blood pressure) and bronchospasms. In their severest form, and if counter-measures are not applied immediately, these responses can be fatal. Despite this potential risk, only 1 patient in 2000 shows any anaphylactic reaction at all.

The other β-lactams can give the same reaction but response to the cephalosporins is often milder; however, patients known to be hypersensitive to penicillin should be considered a potential risk from cephalosporin and carbapenem treatment, though they might be at only a 10% risk of developing symptoms compared with penicillin therapy. Most cephalosporins used are still given by injection and this is unlikely to lead to gastrointestinal problems; however, the increasing number of oral versions, particularly the later, more powerful drugs such as cefixime, have been associated with such severe diarrhoea that courses of treatment have had to be discontinued.

Many of the older antibiotics are labelled as less safe and have been relegated to specific roles. Chloramphenicol has been reserved largely for the treatment of typhoid, with some use in certain cases of meningitis and in the topical treatment of eye infections. The reason for the relegation is that this inhibitor of protein synthesis is perceived as unsafe and it simply does not figure high enough up the therapeutic index to merit use in other than the most severe infections. It is associated with aplastic anaemia and grey baby syndrome. It is also extremely cheap and has become the mainstay of many hospitals in the Third World. When I went to work in Tanzania in 1985, the use of chloramphenicol in the main teaching hospital far outstripped the use of all other antibiotics. The problems that the developed world associated with chloramphenicol use were not apparent in Dar es Salaam. Similarly other sub-Saharan countries use massive amounts of chloramphenicol and the adverse effects are less evident than have been predicted.

Tetracycline has also, to a lesser extent, been relegated. This was the standard broad-spectrum antibiotic of the 1950s and is now used rarely and often only for specific indications. It can cause some kidney or liver damage but only if massive doses are given, so these are rare reactions. It has been shown to be the cause of staining in the teeth of children as the drug, which is yellow, is deposited in the enamel. This type of disfiguration is now considered unacceptable. Because it is used so rarely for acute conditions, it is often considered for continuous therapy of minor infections and young adults with acne have often been prescribed long-term tetracycline therapy. The continuous use of such a broad-spectrum antibiotic can cause problems where indigenous bacteria are keeping other microorganisms at bay. The removal of these microorganisms in the vagina and throat can lead to overgrowth of fungi, in particular with *Candida albicans,* to cause thrush.

Co-trimoxazole, the mixture of trimethoprim and sulphamethoxazole, has caused enormous concern over safety because the benefits of the therapy may not be balanced by risk of adverse reactions. The combination, usually marketed as Septrin, was the subject of a bitter exposé by the press, about six years ago, which eventually led to the government recommending restriction of its use despite previous assurances that it was safe. The combination was marketed together because it was reported that this would have beneficial effects for three reasons. The first is that the two drugs would act synergistically. We have already seen that this is a perceived response that is easy to demonstrate in laboratory tests but is insignificant in the treatment of most clinical infections. The second is that the two drugs are each only capable of inhibiting growth; they are not able to kill the bacteria, but together they produce bacterial death. The results on which this conclusion was based had failed to test the effect of trimethoprim and sulphamethoxazole on bacterial survival in conditions similar to those found in the human body. Bodily fluids produce a metabolic environment that allows trimethoprim, at least, to kill bacteria; this environment was not mirrored in the early experiments conducted to test the action of the drug. The third premise was that the use of two drugs would delay the emergence of antibiotic resistance. This theory was based on the traditional view of the development of chromosomal mutational resistance occurring at a rate of 1 in 10,000,000 – a naïve view bearing in mind that it was being postulated nearly 10 years after the discovery of plasmid-borne resistance genes. Although trimethoprim resistance had not yet been found to be encoded by plasmids, sulphamethoxazole resistance was widely disseminated by plasmids. Indeed this was the major flaw in the third argument for, whether or not you believed that the importation of resistance genes increased development of resistance, half of all bacteria that would be likely to be treated with the combination were *already* resistant to sulphamethoxazole. This meant that, at best, half of all treated infections were effectively being treated by trimethoprim alone. There were lesser arguments that lower concentrations of each drug could be given so that the combination would be safer. In fact the dose of trimethoprim given in the combination raised the concentration of the drug to more than 100 times the inhibitory concentration of the drug.

The scientific arguments for combining the two drugs were weak and largely invalid. The drug had originally been discovered by Burroughs-Wellcome but this had become entangled in a legal issue with another pharmaceutical company, Hoffmann-La Roche. According to the *Sunday Times*, the only compromise to resolve this inconvenience was for the companies to enter into a joint commercial venture. They decided to market trimethoprim in conjunction with a sulphonamide that Hoffman-La Roche had developed, sulphamethoxazole. It had a similar distribution through the body to trimethoprim, so it was supposed that, at most infection sites, the bacteria would be challenged with both drugs. Both companies decided to launch their joint product at the same time, though in the United Kingdom the marketing prowess of Wellcome ensured that sales of Septrin far outstripped that of its twin Bactrim (the Hoffman-La Roche product). This combination was widely used against common infections with apparently successful outcomes, but some clinicians raised questions about the safety of the sulphamethoxazole. It was being blamed for nausea, vomiting and, in the most severe cases, Steven-Johnson syndrome which can sometimes lead to death. It was also determined that the trimethoprim component was relatively safe and free from severe side-effects. If both drugs worked effectively, then the established evaluation of risk could be made according to the therapeutic index, but half the bacteria were already resistant to sulphamethoxazole and even treatment of those that were not was largely by the trimethoprim component because either it reached extraordinarily high concentrations or sulphamethoxazole failed to penetrate the infection site. This raised a very difficult ethical dilemma; how can you prescribe a combination of two drugs when the vast majority of the antibacterial activity was derived from one but almost all the adverse side-effects were derived from the other? This dilemma was strengthened because, when the patent on trimethoprim expired, some minor pharmaceutical companies that specialise in marketing drugs of lapsed patents, nicknamed me-toos, had started to sell trimethoprim on its own. Support groups sprang up to succour those who believed that they were victims of co-trimoxazole therapy and representation was made to the government to have the use of the combination restricted. The then Minister of Health, Mrs Virginia Bottomley, initially disputed any risk to health and then suggested that co-trimoxazole should only be used when trimethoprim alone would be insufficient. There are some incidences where the combination is better than trimethoprim alone and the calculation of the therapeutic index shows an acceptable risk. The most important of these is the treatment of pneumonia caused by the protozoan *Pneumocystis carinii*. This used to be a common cause of lung infection in AIDS patients and was frequently responsible for death although aggressive therapy, including co-trimoxazole, has markedly limited the damage of the organism in AIDS. In this case, the calculation is easy; failure to treat would prolong infection and may be fatal. Co-trimoxazole is often favoured in treatment of gonorrhoea when alternatives have not been available. Trimethoprim alone is ineffective against the causative bacteria. Co-timoxaozole can also be active against *Stenotrophomonas maltophilia* which, as we shall see later, can be the cause of infection in the most severely ill patients in Intensive Therapy Units, and in this case the use of the combination is justified.

The greater awareness of the risks that we might be exposed to with pharmaceuticals has led to severe guidelines to prove the safety of a new drug. The guidelines are usually drawn up by the national regulatory authority in each nation state. In the United Kingdom this is the Commission for the Safety of Medicines, though this power may eventually be transferred to the European Union so that one regulatory body will cover drug usage throughout the continent. In the United States, this role is taken by the FDA, which is the Food and Drug Administration, though it has been nicknamed the Foreign Drug Assassinator by one English Professorial wag, because of their apparent reluctance to grant licences to products from European companies. In fact they are extraordinarily cautious and would cite the example of thalidomide, a drug not passed by them, but which caused so much devastation when passed by our own authority. This caution may be laudable and it is certain that the public has been spared the ravages of some unsuitable pharmaceuticals. However, it also likely that they have been denied some extremely valuable drugs as well. The development of antibiotics has been critically affected by the FDA because the United States accounts for 30% of the total declared antibiotic usage. If a pharmaceutical company wants to license a new drug, it must pass the safety requirements of the FDA. If it does not, it stands no chance of recovering its developmental cost.

It is unlikely that the original antibiotics other than penicillin and the cephalosporins would pass modern safety standards. Certainly the sulphonamides would fail and we would have probably been denied the aminoglycosides gentamicin and streptomycin. This now causes a new problem. As resistance develops, particularly in hospital bacteria, new more resilient species cause infection. The antibiotics required to kill these new pathogens have to be more robust and even more revolutionary in their capabilities; at any rate, they have to be more powerful. This causes conflicting requirements because more powerful antibiotics usually mean less selective drugs. This dilemma is occurring at a time when the FDA has markedly raised its standards of safety and this has partly been responsible for the reduction in new antibiotics currently under safety trials. The FDA demands that all new antibiotics can be used with impunity in infections across a wide spectrum of patients. They demand that they could be given to patients who might undergo long-term treatment in the community with no noticeable adverse effect. In the developed world, the crisis in the management of bacterial infection is in hospital Intensive Care Units in which the patient, unless treated, may well die. In these conditions drugs with a relatively low therapeutic index might be considered acceptable but we are approaching a stage where the therapeutic indices of all new drugs must be of an order that would be acceptable for community use. Therefore a substantial effort of the pharmaceutical industry has been to find antibiotics suitable for community and general use, which they would then target at Intensive Therapy Units. The difficulty with this approach is that any drug considered safe enough for general community use will be used for general infections; it is after all the area where there is the greatest potential return on the developmental costs. This means that antibiotics with lesser safety profiles are not being developed and we are facing difficulties in finding suitable antibiotics to treat intensive therapy patients. A good example of this has been the problems faced by the fluoroquinolone group of antibiotics. The earliest drugs within this group, such as ciprofloxacin and ofloxacin, had a satisfactory safety profile; however, problems emerged as manufacturers tried to increase the power of these drugs by adding yet more active functional groups to the basic nucleus. Some newer drugs were found to be phototoxic; the molecule broke down if exposed to sunlight or ultraviolet. This could occur in the patient as the drug came to the skin surface and sufficient light passed through the skin epidermis. The breakdown of the molecule releases products which may comprise individual aromatic rings, the most serious of which can be benzene. It has been considered that there is no safe minimum concentration of benzene for the body and its presence can place the patient at risk of producing malignant tumours. As far as we know, this breakdown only occurs in the presence of strong light. Ironically the patients who most need new drugs for treatment of hospital-acquired infection need never be placed in strong light, as Intensive Care Units are usually protected from outside light. They are usually lit by fluorescent lights to save money, which do emit ultraviolet rays, but if we really need to use these drugs this could be modified.

Some antibiotics are considered to be potentially carcinogenic or tetatogenic in their own right. This would not be surprising as many of them have action on the DNA of the bacterial cell. The traditional tests to evaluate carcinogenicity is to test compounds by their capability to mutate bacteria. Many antibiotics demonstrate such a mutation capability, such as trimethoprim, fluoroquinolones, sulphonamides, metronidazole, rifampicin. This is not to suggest that they cause tumours in humans, only that they can be mutagenic in bacteria. This raises a problem for the use of some antibiotics, they are less selective in faster-growing human tissue, particularly children. This is a problem that has been identified for some of the fluoroquinolones, which are banned from use in children. This is perhaps prudent but it sometimes masks the optimum therapy option in the face of serious disease. In certain parts of the world, *Salmonella typhi*, the causative organism for typhoid fever, has become resistant to the traditional treatment antibiotics, chloramphenicol, ampicillin and co-trimoxazole. The most promising alternative is ciprofloxacin. It is for this reason that the British Army supplies ciprofloxacin to its troops when they are stationed in the tropics. The use of ciprofloxacin, however, is banned in children who are the group that are most at risk from this infection. Treatment of typhoid in children requires a complete re-evaluation of risk according to the therapeutic index; a much smaller figure could be acceptable for treatment of potentially fatal infections. When my own children were small, if either of them had succumbed to typhoid, there would be very little doubt in my mind that ciprofloxacin would be the favoured option. In my travels in India, there seem to be many others who have made similar risk analyses and favoured the use of ciprofloxacin. It is now probably the drug of choice for typhoid for adults. How many physicians, because of dogma, have not prescribed ciprofloxacin and watched their patients perish? As mentioned earlier, chloramphenicol is largely restricted to the treatment of typhoid because it was considered too toxic to be used for less infection, particularly in children. As far as children are concerned, this may be true for ciprofloxacin; after all, nearly a million people are still killed by this infection around the world. We should not let risk analyses for infections of the urinary tract affect the use of the same drug in life-threatening situations. This is a situation where perhaps a group such as the World Health Organisation should establish guidelines, and the FDA, whose main consideration is for the safety of the American population, should not establish rules for the world, many of whose problems are not mirrored in the USA.

PREGNANCY AND THE NEONATE

Almost all antibiotics are contra-indicated to some extent during pregnancy. Almost all of them can cross the placenta, thus antibacterial management requires reliance on antibiotics that have no adverse effect on the developing foetus. This largely confines the physician to penicillins, some cephalosporins and very rarely some aminoglycosides, though both strepto-mycin and gentamicin has been associated with ototoxicity in some foetuses. Trimethoprim, sulphonamides, tetracyclines and rifampicin have been associ-ated with birth defects in animals, though this may not necessarily be the case in humans, but they have been contra-indicated during pregnancy. Fluoroquinolones have been contra-indicated in pregnancy as well but this is part of a general reluctance to prescribe these drugs in patients with develop-ing tissues. Tetracyclines have the further disadvantage that they can stain the foetal teeth and bone. Sulphonamides are particularly avoided in late pregnancy because they can cause haemolysis in a foetus with a glucose-6-phosphate dehydrogenase deficiency. Rifampicin, although contra-indicated, is most usually employed against tuberculosis and the risks involved are cer-tainly outweighed by the benefits of therapy.

Some of these antibiotics might be contra-indicated if the new mother is nursing her child, because many of them can be passed by lactation. These tend to be the antibiotics that affect the nucleic acid synthesis of bacteria and are likely to be less selective in the very young. Those that are contra-indicated during lactation are trimethoprim, sulphonamides and fluorquinolones and some special precautions should be taken when taking metronidazole. Sulphonamides can also interfere with bilirubin metabolism. Some antibiotics that do not inhibit nucleic acid synthesis such as some cephalosporins and tetracycline are also contra-indicated. The latter can cause the same problems as it does to the unborn foetus. Chloramphenicol also concentrates in breast milk and may subject the neonate to grey baby syndrome. A more insidious adverse effect may result if the nursing mother is given penicillins. These drugs are often in low concentrations in breast milk and are usually totally safe for the neonate; however, the baby may be at risk from sensitisation and this could cause problems later on during subsequent penicillin therapy of the child. Aminoglycosides do not accumulate in breast milk and can often be given without significant risk.

Similar problems exist with the treatment of the neonates themselves. They are one of the serious high-risk groups for infection, particularly if they are premature. Again the nucleic acid synthesis inhibitors such as trimetho-prim, fluoroquinolones, sulphonamides, metronidazole are usually inappro-priate. Aminoglycosides are often inappropriate because of the risk of ototoxicity unless the risk outweighs the disadvantages. Chloramphenicol and tetracycline are considered unwise for the same reasons that were asso-ciated with breast milk. Rifampicin is also considered unwise in the very young. This significantly reduces the options often to pencillins, cephalosporins and macrolides such as erythromycin. Significant resistance problems to this limited group of antibiotics puts the newly born, especially those in special-care baby units, at very great risk.

INTERACTIONS WITH OTHER DRUGS

A question that is often asked is, if I am taking an antibiotic can I drink alcohol? The answer is not an easy one and is, really, it depends. Metronidazole, or Flagyl©, specifically states that alcohol should not be taken during therapy. It is likely to cause nausea, even vomiting, as well as a drop in blood pressure. The taking of alcohol is specifically not recommended with certain cephalosporins because the same effects are found. In these drugs the presence of alcohol inhibits the enzymes that break down the antibiotic to its active components. The use of alcohol is not specifically forbidden with most other antibiotics but, of course, it should always be used in moderation.

An interaction, far less often considered, is that with the contraceptive pill. Rifampicin is specifically contra-indicated because it induces enzymes that reduce the effective absorption of the pill; however, there is a much more general consideration. Any antibiotic that produces diarrhoea is rendering co-administration of oral contraception ineffective. While the body is trying to expel the contents of the gastro intestinal tract, any concurrent therapy has little or no chance of absorption. This may be particularly evident with some of the penicillins where diarrhoea may be experienced by 20% of patients. This produces an unacceptable risk of pregnancy for any form of contraception so those on antibiotic therapy as well as the contraceptive pill should consider mechanical forms of birth control for the month surrounding therapy. The contraceptive pill works by maintaining a sufficient, but low, dose of hormone in the body. Any disruption to this could introduce the risk of pregnancy until the next period. According to the *Monthly Index of Medical Specialities* (MIMS), the antibiotics that could cause gastrointestinal problems besides penicillin are cephalosporins, tetracyclines, trimethoprim, fluoroquinolones, sulphonamides, metronidazole and rifampcin. This list is not complete but merely serves to show what precautions need to be observed. The extent of the risk can only be assessed by the individual patient.

If the patient is taking something apparently innocuous he or she may still be compromising the antibiotic therapy. This may be evident in patients drinking milk while taking tetracycline, which can annihilate the action of the drug; similarly, if the patient takes some therapy that provides either magnesium, aluminium or iron salts. This may be an iron supplement for a patient showing anaemia though this is likely to be under medical control. A significant proportion of the population suffer from heartburn and self-administer antacid buffers that neutralise the stomach acid. Recently they also have the alternative to self-administer an H_2-blocker such as Zantac. All these remedies may interact with fluoroquinolones whose absorption and antibacterial activities are very susceptible to the concentration of any substance that provides monovalent or divalent metal ions. The fluoroquinolone can *chelate* the metal ion which can radically affect its absorption. Thus patients who suffer from heartburn should refrain when taking fluoroquinolone therapy or at least not take these remedies within a two-hour period just before or just after the antibacterial. Multivitamins also contain metal cations and they also interfere with the absorption of fluoroquinolones, so should be avoided during therapy.

As increasing number of the population, especially children, in the developing world suffer from asthma, and many are given theophylline. This is a stimulant drug, derived from tea, that acts as a bronchodilator, opening the restricted airway passages, which alleviates the manifestation of asthma to restrict passage of these airways. The problem of respiratory disease is that it is often not known whether it is caused by an inflammatory response or by bacterial infection. Indeed, restriction of the airway passages in the lung can lead to bacterial infection. We have an elaborate system of cilia in our trachea (windpipe); these are small hair-like tentacles that beat in unison and physically remove invading bacteria from the lungs. Asthma restricts their movement which leaves the patient prone to infection. Unfortunately theophylline levels are affected by a number of antibiotics given specifically against respiratory infection. This is a particular problem for patients given fluoroquinolones; the increase in theophylline is especially high with enoxacin giving an increase of 111%, compared with the more modest increases of 23% for ciprofloxacin and 12% for ofloxacin. The 4-oxo metabolite of the piperizine ring of the fluoroquinolone is thought to interfere with some liver enzymes, particularly hepatic cytochrome P450, and reduces theophylline clearance. Asthmatics; whose bacterial infections are treated with fluoroquinolones should be warned of the dangers and, particularly in the case of enoxacin, the dose of theophylline should be halved. This is not necessary with the other fluorquinolones but patients might wish to have their theophylline levels monitored while on fluoroquinolone therapy. Rather less seriously, the fluoroquinolones also delay the clearance of caffeine and enoxacin can increase its absorption and delay its clearance significantly. This is an interaction that is unlikely to be noticed by the patient.

Many of the older drugs, such as chloramphenicol and sulphonamides, have other interactions. These drug are cleared by the liver and they can interfere with some liver enzymes and can compete with other pharmaceuticals. This can not only increase the clearance of the antibacterial, thus reducing their efficiency, but it can also prolong the presence of the competing drugs. This is particularly noticeable with phenytoin, a drug given for heart arryhthmias, which can reach dangerous levels if not cleared. This may lead to ataxia, impairment of the ability to co-ordinate the muscle movement, and the patient may have a tendency to overshoot an object when reaching for it or, in a more serious form, have difficulty in walking or in maintaining balance. These two drugs can, to a lesser extent, interact with the depressant phenobarbitone, which used to be used widely for insomnia. Although care should be taken with these drugs, there are suitable alternatives and, as they have significant adverse effects in their own right, consideration of their interactions with other drugs should, for the most part, be academic.

The most contentious interaction at the moment is with the carbapenem, imipenem. This group of β-lactam antibiotics is considered, by some, to be the final defence against serious hospital infection. Imipenem was the first of this group of antibiotics to obtain a licence. Unfortunately it has a significant disadvantage; it is metabolised by dipeptidases and dehydropetidases in the kidneys and it cannot be administered unless the activities of these enzymes are repressed. Therefore, imipenem is administered with a peptidase inhibitor, cilastatin sodium, to limit the activity of these enzymes. This meant that the drug could not be given to children under the age of 12, which meant that this particularly vulnerable group could not benefit from this most powerful of antibiotics. There were also suggestions that children might suffer fits if given the combination of imipenem/cilastatin. A more recent carbapenem, meropenem, has overcome the problem of the kidney enzymes and can be given without an inhibitor, which ensured that it could also be given to children, and indeed it is recommended for meningitis. In my own locality around Edinburgh, this advantage has ensured that the later carbapenem has completely overtaken the first drug. This is unusual, as generally the first drug to be launched in a series of drugs has such an advantage in sales that its lead is never seriously challenged. Only when the subsequent drug is seen to have such a significant advantage can it overturn this domination of the market.

ANTIBIOTIC INTERACTIONS
Antibiotics can have interactions with each other. The bactericidal antibiotics do not simply interfere with some process in the cell which kills the bacterium. Instead, the antibiotic usually initiates a death response in the bacterium that is complex and, for many antibiotics, is hardly understood. It usually requires the synthesis of proteins, without which the cell will not die. A significant number of the antibiotics, which are just bacteriostatic in their action, are inhibitors of bacterial protein synthesis. Thus if they are added together with an antibiotic that initiates a death response, the protein synthesis inhibitor prevents the manufacture of those proteins essential to promote bacterial death. The result is that the bacteriostatic antibiotic is antagonistic to the action of the bactericidal drug and the bacteria do not die. The usual recommendation is that bacteriostatic antibiotics, which are inhibitors of protein synthesis, should not be co-administered, – at least not given before or concurrently. It is conceivable that if the bacteriostatic drug is given after the bactericidal antibiotic, it is too late to influence the death cascade. The antibiotics whose death cascade is most affected by the bacteriostatic protein

synthesis inhibitors are the β-lactams and perhaps the aminoglycosides. The effect on the latter is ironical because these antibiotics are supposed to be protein synthesis inhibitors in their own right. The β-lactams and aminogly-cosides are usually contra-indicated with protein synthesis inhibitors but it may be that this caution is theoretical rather than real. About 16 years ago, when my father had suspected bacterial meningitis, amoxycillin and chlo-ramphenicol were prescribed. In emergency cases such as meningitis, it is essential to provide sufficient antibiotic cover to guarantee that the causative bacterium is sensitive to at least one of the drugs given. Some of the bacteria that could cause meningitis are resistant to amoxycillin, which is particularly unlucky because this is the one antibiotic that probably penetrates the meninges most effectively. Chloramphenicol also penetrates but, unlike amoxycillin, is bacteriostatic; it is given to inhibit those bacteria that are resistant to amoxycillin. Does the presence of chloramphenicol compromise the effect of amoxycillin in those bacteria that are amoxycillin-sensitive? It does not seem to, but the ability of amoxycillin to kill the bacterium is proba-bly not important in meningitis. In a patient with an effective immune system, the ability just to prevent growth is probably all that is required. In severe infection, caused by an aggressive hospital pathogen, the bactericidal activity of the antibiotic is usually an important asset. There is often little place for a bacteriostatic antibiotic so the risk of adding the two together is not great.

The co-administration of two bactericidal antibiotics is supposed to be synergistic; the effect of two combined is reputed to be greater than the expected sum of the individual responses. This ability of an antibiotic to kill is a surprisingly difficult parameter to quantify and it is almost impossible to sum the killing capabilities of two antibiotics. At the very least, two bacterici-dal antibiotics are not antagonistic and each antibiotic is unlikely to interfere significantly with the effects of the other.

There are some antibiotics that seem to have a bactericidal effect that is partially dependent on protein synthesis and the rest is not. The fluoro-quinolones are able to kill sensitive bacteria rapidly. If they are co-adminis-tered with a bacteriostatic antibiotic, there is a significant reduction in the killing effect. However, unlike the other bactericidal antibiotics, there is not a complete reduction, perhaps just 50%. This means that if co-administered with a bacteriostatic antibiotic, the killing action of fluoroquinolones might not be totally annihilated. In general, mixing bacteriostatic and bactericidal antibiotics should be avoided.

Bactericidal antibiotics might never seem detrimental in their own right. It is sometimes detrimental if the bacterium is lysed. In general, most bactericidal antibiotics lyse the bacterial cell at the end of the death cascade, and the contents of the bacterium are released. Some bacteria produce endo-toxins that are relatively innocuous when inside the cell; the main purpose of the production is not to kill the human host. However, if an antibiotic aggressively lyses the cell and releases large amounts of toxin, the patient may go into shock.

PERSISTENT DRUGS

The trend in antibiotic development is to produce drugs that have longer half-lives. This has enormous advantages both in the cost of administration and also in patient compliance. Many antibiotics are now capable of being given once a day instead of two or four times a day as in the past. This is achieved by delay of the clearance of the antibiotic from the body through either the kidneys or the liver. If the antibiotic has any significant adverse effect it will mean that this capability also has a longer half-life. It may place the patient at a greater disadvantage. Traditional antibiotic administration is based on peaks and rapid clearance. This means that the antibiotic is removed quickly and for much of the time, particularly just before redosing, the concentration of antibiotic will be considerably lowered. In the longer half-life drugs, the antibiotic remains at a higher concentration for a much longer period of time. This has obvious advantages for therapy but if the antibiotic is toxic, especially if this toxicity is cumulative, then continuous exposure to high levels of drug could be detrimental. It is even more important with these newer longer half-life antibiotics that care is taken to be vigilant for the appearance of adverse effects.

QUALITY CONTROL

As with all pharmaceuticals, the presence of other components in antibiotic preparations may be a cause of concern. In the United Kingdom and in many of the countries of western Europe, the manufacture of antibiotics is strictly controlled but many of the antibiotics that we consume are not actually manufactured either in this country or in the European Union. The selling of an antibiotic is a commercial business and the wholesaler will be attracted to wherever he can obtain the cheapest deal. For this reason, a significant proportion of antibiotics, particularly the generic out-of-patent compounds, have been imported, often from eastern Europe. This is not to say that these generic compounds are unsafe; many of them are undoubtedly made under strict conditions of hygiene and the chemical content is as pure as possible. However, the origin of these antibiotics may not be eastern Europe and may be much further afield. Antibiotic manufacture in parts of the developing world is far less rigorously controlled.

Ciprofloxacin is manufactured in India by more than 80 companies, and not one of them pays a penny to the patent holder, the pharmaceutical giant, Bayer. This antibiotic is not difficult to synthesise but a large ethical pharmaceutical company like Bayer has to ensure that the compound is as pure as possible and that all impurities are extracted. They also guarantee that the concentration of the active component in the preparation is, within defined limits, what is stated on the packet. Patent infringement with "pirated" antibiotics is often considered by consumers simply to be unfortunate for the company that has had its product poached. It is, however, much more insidious because many of the companies that manufacture ciprofloxacin do not have the reputation of a multinational to preserve and their products are often, at the very least, inadequate and, at worst, positively harmful. Within that enormous spectrum of pirated ciprofloxacin products, there are some that have an activity equivalent to that manufactured by the patent holder. In 1989, I purchased some local ciprofloxacin in India and analysed it. It was as active as the Bayer product obtainable in the United Kingdom, it had the same killing power and had almost the same chemical spectrum. There was, however, a spectrum line that was not present in the Bayer product. We never analysed this further but it may have been a contaminating chemical. This compound was made by the Rambaxy Pharmaceutical Company, regarded by many Indians as one of the more responsible chemical companies. Others scientists, including those from Bayer, have analysed ciprofloxacin manufactured by lesser companies and purchased in India. They have found that what is called ciprofloxacin varies from near perfect copies, such as that from the Rambaxy Pharmaceutical Company, to products that have no active ciprofloxacin at all. In between, there are compounds that have antibacterial activity but include another, and usually less effective, antibiotic, as well as compounds that do possess ciprofloxacin but far lower concentrations than listed on the packet.

Any pharmaceutical that is an inadequate imitation of what it purports to be is a cynical attempt to extort money with no regard for the suffering that it is going to cause. A single tablet of "ciprofloxacin" represents more than a day's wages for some agricultural workers in India. Surprisingly, after the near perfect copies, the next best preparations could be considered to be those that contain no active antibiotic. This may seem very harsh on the hapless patient, who will enjoy no antibiotic protection and may die if the infection is serious. It will mean, however, that the pathogenic bacteria will not be challenged with inadequate doses of ciprofloxacin, a situation that will promote the development of resistance, and that subsequent patients afflicted with this bacterium may still be treatable with pure ciprofloxacin. The very worst drugs are the ones that contain small quantities of antibiotic, whether that be the antibiotic stated on the label or some other compound. These are storing problems for future patients and, as we shall see in a later chapter on antibiotic resistance in the developing world, they have caused problems which are already with us. If the preparation is not mainly composed of active antibiotic, what is the white powder that the patient is swallowing or having injected, and this, as a consumer, would be my greatest concern. In Europe and north America, there are good lines of communication to identify the emergence of adverse effects. Poor communications is one of the definitions of a developing country and it is virtually impossible for any manufacture to have reliable feedback about adverse effects, even if they sought it.

THE PHASES OF DEVELOPMENT
The dilemma
The safety profiles of antibiotics are dominating the development of the drugs. The FDA is the effective guardian of antibiotic safety and it requires demonstration that a new antibiotic is a significant improvement over those that we already possess. In simple terms, it means that the drug usually has to be more reactive chemically than its predecessors. The necessity to produce more active compounds comes from the selection of bacteria resistant to the previous generations. If therapy with the previous generations of antibiotic removes a hospital pathogen such as *Klebsiella pneumonia* and provides the environment for the multi-resistant *Acinetobacter baumannii* to proliferate, the next generation of antibiotics will have to be able to cope with the new pathogen. The new pathogen is often already multi-resistant and very difficult to treat. It requires a more active chemical compound and these, almost by definition, are likely to be less selective in their action than the previous generation of drugs. On the other hand, the FDA is insisting that safety profiles of the current antibiotics under development are higher than the previous generations. So they are looking for more active compounds with greater selectivity; indeed they appear to require that all antibiotics could be used for the treatment of minor as well as life-threatening conditions. This is often paradoxical and the problem of phototoxic fluoroquinolones demonstrates the problems that have emerged. This stance means that a number of antibiotics will not be permitted to be marketed because they possess an insufficient safety profile, although their therapeutic indices might suggest that the risk of treatment might be outweighed by the possible life-saving potential. The pharmaceutical industry is moving away from the development of antibiotics, as the return may not repay the investment. It is, of course, improper to permit an antibiotic to be used if it has known serious adverse effects; however, in a situation where the patient is at serious risk of death, the safety profile of an antibiotic is little comfort to the family. The aminoglycosides, which have been responsible for the saving of literally millions of lives, would have difficulty in passing current safety regulations. Nowadays, no responsible clinician would use these antibiotics outside the hospital and usually only to control serious infection. We have already hit a crisis of insufficient antibiotics to treat hospital infection, so more drugs are necessary. It is perhaps time that the controlling authorities permitted clinicians, with well-defined guidelines, access to antibiotics that may be less safe but could save the lives of severely compromised patients.

The problem is, of course, litigation. The legal attack is on the medical practitioner, the licensing authority and most of all, perhaps because they are perceived as having most money, the pharmaceutical companies. Massive medico-legal settlements were conceived in the United States but are already starting in Europe. All negligence should be open to legal attack, but the clinician is also obliged to save life by all means available to him or her. In the face of a dwindling antibiotic resource, we certainly need more drugs and we probably need some of the ones that have poorer safety profiles. If we do not, we shall soon have to re-evaluate our medical capabilities in the future. In particular, we shall to reconsider the risks of transplantation surgery. If a surgeon cannot be sure that the antibiotics are available to cover the inevitable infection that would follow the immuno-suppression for this type of procedure, surely he or she could be considered negligent.

The lack of enthusiasm of pharmaceutical companies to develop antibiotics has already reduced the flow of new drugs. Indeed, there have been no new classes of antibiotics for at least 10 years and the situation does not appear to be improving. In 10 years' time, there are likely to be virtually no new antibiotics and while resistance to the current drugs increases, a crisis in the management of hospital infection appears likely. What will happen then? For an answer, I think that we might take our cue from the AIDS crisis of the late-1980s. At that time, we were facing the potential of a global epidemic of an infectious disease, which in its later stages can be excruciatingly painful and predictably fatal. There was also no obvious cure. The Wellcome Foundation had, many decades earlier, developed a drug called zidovudine or azidothymidine (AZT). AZT was promoted because it interferes with the replication of the Human Immunodeficiency Virus and rapidly conceived clinical trials demonstrated that it could prolong life significantly in some patients and delay the onset of full-blown AIDS in persons with no symptoms. This compound was not developed as an anti-virus agent at all. It was originally designed as an anti-bacterial drug by George Hitchings, at the same time that he was developing trimethoprim. Trimethoprim was selective as an antibacterial, but AZT, although very active against bacteria, was not selective and had side-effects that were incompatible with a drug that might be used to treat common bacterial infections. When a much more serious, life-threatening infection emerged, AZT could be used with impunity even though its toxic side-effects even now preclude its use in many cases. The pharmaceutical industry is a commercial business and prices compound according to market forces. AZT was extremely expensive when first marketed and the success of the therapy raised the financial fortunes of the company.

It is likely that in 10 years' time, our forecasted crisis in antibiotic options to treat hospital infection may force the licensing authorities to relax their own rules for antibiotic safety. Many companies currently possess drugs that, like AZT, would be ready to enter clinical study if such a relaxation occurred. It would be likely that the first company to obtain a marketing licence would be able to charge whatever they wished and would be set to make a massive return. Once competition occurred with other drugs, the price would fall.

As the industry is currently structured and the current guidelines laid down by the regulatory bodies are adhered to, it would be difficult to see how pharmaceutical companies can be persuaded to continue to search for new antibiotics. It will cost them up to $500 million to bring a brand new antibiotic to market; if the directors make the wrong decision and the antibiotic fails late in development, even the most robust company would reel from the financial consequences.

A socialist might consider that this kind of drug development should be taken out of the control of large, multinational chemical companies. This was the situation in eastern Europe before the removal of communism and it can be stated that no new antibiotics were ever developed by these state-run monopolies. In fact, they were simply churning out antibiotics discovered in the west and often trying to undercut the market in the west. There appears to be no alternative other than the free-market development of antibiotics but if we wish to have a continuing supply of drugs we should evaluate what we really want from the industry.

7

Intensive farming – the use of antibiotics in animals and the consquences for man

During the Second World War, the island fortress of Great Britain stood alone against the Nazi threat for over a year between the evacuation of Dunkirk in the summer of 1940 and the attack on the American Pacific fleet at Pearl Harbour in late 1941. The island was besieged as many castles and towns had been in the Middle Ages, though, unlike medieval sieges, the assailants were not massive armies but U-boats. However, the principle was the same, to try to starve the besieged into submission. This island of around 45 million people could not feed itself and so it had to import food from outside. The U-boats sank hundreds of thousands of tons of shipping which were simply carrying food.

Before the war, the imperial powers, particularly Great Britain, had used their empires to supply the mother country. The land masses that they controlled were vast and often relatively under-populated. Many colonialists had set up large farms whose sole purpose was to produce food for the imperial master. This agriculture was inefficient and labour was cheap. Farming in Great Britain was also ineffective and often conducted haphazardly. The government had subsidised much of the production and the yields were low. The siege of the war suddenly revealed the liabilities of inadequate home production and reliance on colonial supplies. Food rationing and careful distribution of available supplies prevented actual starvation but it has been considered that this might have resulted from luck rather than judgement. Food rationing did not disappear at the end of the war and persisted for more than five years. On the other hand, the United States, with its almost limitless available agricultural land, had easily been able to fulfil its own needs.

The threat of starvation during the war and then the delayed removal of food rationing after the war convinced the British government that the island must develop its own agriculture so that it could supply sufficient food for the whole population. Thus from the early 1950s, the government instigated measures to improve the efficiency of the home agricultural base. The fertility of the soil could not sustain the number of crops required by the whole population so fertilisers were added to boost yields and shorten or remove the time required for the land to recover from one crop to the next. The land area could not sustain the domestic herds required to supply sufficient meat to the population, so new methods were adopted to rear cattle and pigs. Instead of allowing them to roam in large fields, letting them forage on their own, much greater numbers could be reared if the animals were confined and presented at regular intervals with feed of a higher nutritional value. This would mean faster growth of the animals and better yields for the investment.

This policy was found to be particularly effective in the rearing of calves and pigs. It was also found to be very productive in the rearing of poultry. Chickens had traditionally been reared in farmyards where they were permitted to roam freely, but this limited egg and meat production and made it expensive. The solution was to confine large numbers of birds in closely packed stalls and feed them at regular intervals. For the first time, farm animals were suddenly collected in large numbers for long periods of time. This would open up the opportunity for the spread of infection.

Infection in agricultural animals was not new but usually it was limited and containable. If the animals within a range-reared herd became infected, a vigilant farmer could limit the damage by isolating the infected animals; however, intensive farming meant that animals were more likely to infect one another and, instead of confining the infection, the close proximity of the animals could mean that a whole herd could become infected. The consequences of such widespread infection could be catastrophic. A virulent infection might kill large numbers of animals but the financial consequences could be felt even by relatively minor infections. Many bacterial infections reduced appetite, which would delay weight increase and reduce muscle gain. This would lengthen the time taken to prepare the animal for slaughter or reduce egg or milk production. Farmers needed a means of controlling infection in high-intensity animal production and antibiotics were the solution.

In the 1950s, it was found that if antibiotics were fed to animals they grew more quickly. This had been largely by accident because chickens fed with a vitamin B12 supplement were healthier and put on more weight than chickens that had not been given the supplement. It was subsequently found that the vitamin was not itself the cause of the dramatic improvement but rather it had been contaminated with tetracycline and it was the antibiotic that was responsible. This initiated widespread use of antibiotics as food additives, resulting in some quite staggering improvements. They proved particularly effective in intensively reared animals, poultry and pigs but when cattle farming became intensified, mainly for veal production, substantial improvements in weight gain were also found if antibiotics were added to

FDA approved		UK approved
Bacitracin	Oleandomycin	Avilamycin
Bambermycins	Oxytetracycline	Avoparcin
Chlorotetracycline	Penicillin	Bacitracin
Erythromycin	Salinomycin	Flavophospholipol
Hygromycin B	Streptomycin	Monensin
Lasalocid	Tylosin	Salinomycin
Lincomycin	Virginiamycin	Spiramycin
Monensin	Arsencials	Tylosin
Neomycin	Carbadox	Virginiamycin
Novobiocin	Nitrofurans	
Nystatin	Sulphonamides	

DuPont & Steele, 1987. Reviews of Infectious Diseases. Veterinary Medicines Directorate

Antimicrobials used as feed-additives

their feed. Laboratory experiments do demonstrate that animals raised in a bacteria-free environment do grow substantially faster than those exposed to normal bacteria, even those harmless commensal bacteria that all animals normally carry as passengers throughout their lives. Coates demonstrated that conventionally raised animals achieved weight gains similar to those reared in bacteria-free environments, provided that antibiotics, usually given at low levels, were added continuously to their feed.

It can be surmised that the bacteria in the gut of animals are competitors for the same nutrients. In humans, the bacteria in the gut are mostly concentrated at the end intestinal tract and these bacteria have to survive on our waste products. In animals that are normally ruminants, bacteria become an essential part of the digestive process. Ruminants largely survive on grass, but plant cells have a rigid cell wall composed mainly of cellulose. The mammalian digestive tract does not possess the correct enzymes to break down this cellulose and so release the contents of the cell, which is the main nutrient benefit of grass. Ruminants have had to develop different methods to destroy cellulose. Cattle have entered into a symbiotic relationship with many bacteria. Unlike ours, the ruminant stomach is divided into four compartments. The first, known as the rumen, is where most fermentation occurs. The rumen contains both bacteria and protozoa which release cellulose-digesting enzymes to break down the cellulose of the cell wall, mainly into short-chain fatty acids as well as other compounds. These fatty acids and other nutrients enter the bloodstream through the rumen and, for cattle, may provide up to 70% of the animal's energy supply. It is the reason why cattle produce large amounts of gas in the form of carbon dioxide and methane. After partial fermentation in the rumen, some of the remaining fibrous material is formed into small round masses called cuds, and brought back to the mouth where they are chewed. This is when cattle are seen "chewing the cud" which is a mechanical grinding of the cells in the teeth to further the release of the cellular nutrients. The chewed cud is re-swallowed and this time the food passes into the second chamber, the reticulum, where some additional fermentation takes place, and then into the third chamber, or omasum, where most of the water is reabsorbed. From the omasum the food passes into the fourth chamber, the abomasum, or true stomach, where ordinary digestion occurs.

As the bacteria break down the cellulose, they are the first to have access to the nutrients that the lysed cells are releasing. Thus the ovine stomachs are not just for feeding the host but also the symbiotic bacteria. The nutrient loss to an individual bacterium is negligible but to many billion it becomes considerable and the host animal has to eat far more than is required for its own individual needs. Intensively reared animals are not usually fed grass but food pellets with a much higher energy content. These pellets, often high in protein, do not require the pre-release of cell-bound nutrients and thus the bacteria in the early stomach chambers are no longer required. Indeed, they are detrimental because these bacteria can release enzymes that break down protein. The presence of antibiotics in the feed, even in sub-inhibitory concentrations, will hinder the metabolism of the bacteria in the rumen and reticulum. This will allow more nutrients to pass to the omasum and particularly the abomasum. The antibiotics that work best as growth promoters are those that are poorly absorbed from the gut and thus remain virtually intact and sustain their concentration throughout the digestive tract.

There is an alternative view of the effect of antibiotics, which I think is less plausible, which is that the presence of these drugs prevents a series of "sub-clinical" diseases to which all mammals are victim during the growing stages. These "sub-clinical" diseases would cause temporary periods of poor nutrition absorption and halt development. Certainly in areas of poor nutrition, growth development is retarded but reared animals in the developed world now receive feeds of the very highest and most concentrated nutritional value, which makes this argument less cogent. This debate could be resolved by observing the effect of continuous antibiotics administration on the growth of natural carnivores. These animals do not have the bacteria-ridden early stomach chambers of ruminants so growth promotion with antibiotics would probably be provided only by the removal of "sub-clinical diseases". We do not farm natural carnivores for food so it is impossible to deliberate the exact role of antibiotics in growth promotion.

A variety of substances have been used for growth promotion. In the United States, hormone implants have been used to increase weight gain although the European Union has banned these implants in its own cattle and, since 1988, prohibited the importation of American beef because of concerns that hormone residues in the meat would have deleterious health effects. In Europe, more use has been made of chemicals that inhibit bacteria. These have not always been antibiotics as copper salts and arsenicals have also been used in low quantities. Copper could give a 5% increase in weight gain when given in a variety of salts, mainly as carbonates, sulphates or oxides and at concentrations as low as 100 parts per million. It was found to be even more effective if given as organic phosphate, copper citrate, when it could be administered in even lower concentrations, although environmentalists have been concerned about the accumulation of copper from animal wastes on the land. Arsenicals have been known to be effective antibacterials since Ehrlich but their lack of selectivity and accumulation in the meat of the slaughtered animal has restricted their use.

When antibiotics were first used as growth promoters, there was a free-for-all. Any antibiotic that was available for human medicine was tried as a food additive. Most seemed to work effectively and as farming became more intensive in the 1950s, so the concentrations of antibiotics used increased as well. They were producing greater than 10% increases in weight gain in pigs, poultry and cattle when given at concentrations as low as 40 parts per million. In terms of the economics of overall production, this results in a massive saving. When I conducted a project with a major chicken grower on the outskirts of Edinburgh, I was told that each bird was expected to take 52 days from hatching until it became ready for slaughter. The profit margins were so low that if the bird was not ready until the 55th day, the producer was making a loss. The use of antibiotics ensures that deadlines such as these can be met. They not only ensure a more efficient conversion of nutrient into weight gain but they also reduce the waste products, particularly faecal waste and nitrogen. In financial terms, the return for the use of antibiotic growth promoters has been estimated as 7-fold and, if their use were banned for the whole of the European Union, the annual cost of meat would rise by nearly £1 billion.

Despite these clear financial benefits, the use of antibiotics in animal husbandry increasingly engenders criticism. There are essentially two problems associated with antibiotic use; the first in the antibiotic residues left after administration, and the second is the emergence of resistance either in pathogenic bacteria or which may be transferred to pathogenic bacteria. The former can either be transferred directly to the human consumer who would then unwittingly ingest low concentrations of antibiotics or these could be released into the environment, particularly onto agricultural land so would be taken up by other animals. It may even lead to problems with the latter.

Some antibiotics, including penicillins, are extremely labile, and they are often so labile that they decompose while the animal is being treated and residues are not a problem; however, antibiotics may undergo some protection in certain tissues and decompose much more slowly in certain animal organs. Other antibiotics can persist for long periods and these are usually the synthetic antibiotic chemicals. In either case, antibiotics make up the majority of all residues found in meat. These are usually low but can sometimes be significant. In order to reduce their impact a series of guidelines have been developed to try to ensure that by the time the meat reaches the consumer, the antibiotic has dissipated sufficiently to reach acceptably low levels. If guidelines are rigidly adhered to, the level of antibiotic remaining should be less than 1% of that administered. These guidelines consist of a withdrawal period, measured in days, where the animal should not be given antibiotics prior to slaughter. The time taken for the antibiotic to dissipate can vary considerably; in terrestrial animals this can be quite straightforward and would vary little from one batch of animals to another. On the other hand, in marine farms, the dissipation depends on the salinity of the water and particularly on its temperature. Variations in temperature can alter the withdrawal period by many days. Antibiotics used as growth promoters usually do not fall within these guidelines because the level of chemical used is considered to be too low and, anyhow, they are usually drugs that are poorly absorbed from the intestinal tract.

Adherence to the guidelines requires commitment and an understanding of the dangers of failure to comply. In the United Kingdom the Royal Pharmaceutical Society is responsible for ensuring that guidelines are respected; they have the power to monitor residues in animal carcasses and impose fines for breaches of the guidelines. The main reason for high residues may be the inconvenience of altering the feeding pattern of animals; this may also be expensive as feed lacking the antibiotic has to be set aside and used just before the animals go to market. The farmer may not have finally decided when to submit his animals to market, as it may depend on fluctuating meat prices which can vary considerably. A switch to non-medicated feed is almost a commitment to slaughter. Sometimes, the use of antibiotics for medication may largely supersede their use as growth promoters; the method by which animals are normally given antibiotics means that much more antibiotic is used than if the equivalent number of human patients was being treated. Unlike antibiotic administration to patients, when an infected animal is found in a herd, the whole herd often receives treatment because the healthy animals are given antibiotics as a prophylactic precaution. As animals are incapable of taking infection control measures, this is often an essential measure if the condition of the animals is to be preserved. Once such medication is administered, it can become difficult to estimate the withdrawal procedure and estimate the drug-free days required. This is where the essential problem lies.

The withdrawal period is an estimate and not based on any chemical monitoring of the antibiotic levels so it is based on the *average* antibiotic dissipation of the antibiotic, which may vary considerably from one animal to another. This calculation has taken the results of toxicity studies, which have been performed on animals used for human consumption, to establish the Acceptable Daily Intake (ADI), which is the daily intake that, during the animal's lifetime, will have no detrimental effect on the consumer's health. This is an essential difficulty with the monitoring procedure. It is designed to prevent toxic levels of antibiotic reaching the consumer; however, the greatest threat of antibiotic use is not usually toxic damage from the chemical interaction with human cells but rather from the selection of resistant bacteria, either in the animal itself or later once the antibiotic passes to Man, which will overcome the clinical use of the same or related antibiotics. The concentrations required to select these resistance mutants are very much lower than those normally required for toxicity.

There is perhaps one exception to this, because most of the adverse reactions involve the use of pencillins and this is usually in dairy products. This drug is often used for the treatment of bovine mastitis which can seriously reduce milk production. Penicillin is often given by intramammary infusion. Penicillin has already been seen to cause adverse reaction in patients who are hypersensitive to the antibiotic after they have been "sensitised" by a previous course. The most common reaction to penicillin residues is allergic but, perhaps because of the low levels, the responses are usually fairly mild, with dermatitis the most common reaction. The severest reaction to penicillin hypersensitivity is anaphylactic shock and no residue-induced responses had ever been documented by 1993.

Some observers reassuringly consider that the antibiotic levels are simply too low to select resistance in the resident bacteria of the intestinal tract; however, such considerations come from the popular view that resistance derives from the heavy consumption of antibiotics. Resistance development is no respecter of the host; in the chapter on resistance in the developing world we shall see how low concentrations of antibiotics are the most potent selection environment ever devised. Indeed, the levels of resistance in intestinal bacteria from terrestrial farm animals do not bear out this reassurance.

Bacterial resistance to antibiotics used in veterinary medicine and as feed additives is disturbingly high. In the 1960s, new antibiotics were regularly introduced into clinical medicine and, almost concurrently, the same antibiotics were introduced for growth promotion and for medication in animals. The intensive farming procedures, introduced over the previous decade, ensured that infection was particularly high amongst animals stocked together in high-density pens. At the time, the only disease consid-

Factory farming – promoting the spread of antibiotic resistance

ered to be of any significance in cattle was the virus-borne foot-and-mouth disease. The tastes of the public were becoming more adventurous and, in the days before animal welfare became a political issue, veal was becoming popular. Veal had always been a relatively expensive product in comparison with the subsided beef production; however, if the animals could be penned during their short lives, the costs could be reduced considerably. Veal production expanded rapidly and so did contamination. The animals were soon to have a much greater than average carriage of the pathogen *Salmonella typhimurium*, a bacterium that can cause severe intestinal disease in humans. A specific strain of *S. typhimurium*, called type 127, was identified by E.S. Anderson and colleagues at the Enteric Pathogen Reference laboratory at the Central Public Health Laboratory in Colindale, London. It meant that now the consumer was suddenly threatened by infection from the meat that they were consuming, which seems a naïve view nowadays with repeated threats to the safety of agricultural products but, 35 years ago, was unexpected. It became essential to cook beef products thoroughly and, more important, to separate uncooked products from cooked within the refrigerator or food preparation areas. Anderson found that, as each new antibiotic was introduced in the 1960s, the type 127 *Salmonella* acquired resistance to it within two years. This forced farmers and vets to use the most newly introduced antibiotic in order to control this infection. The epidemic continued to increase and was traced to a source supplier of veal cattle, who was distributing animals throughout the United Kingdom. We did not possess the legislation to control the spread of this infection problem and it was not solved by the introduction of legislative control but rather when the owner of the veal distribution company was killed in a car crash and his business subsequently failed. The numbers of type 127 *Salmonella* dropped considerably and by the beginning of the 1970s, the epidemic was effectively over.

Anderson's results were causing alarm amongst the consumer public and the press. He argued that there was no control over the use of antibiotics in animal husbandry and that we were in danger of losing antibiotics for clinical use under the mass of resistant bacteria that was being generated in agriculture. A Royal Commission was set up under the chairmanship of Professor Sir Michael Swann. The Swann report, delivered in 1969, recognised this danger and deliberated that growth promoter antibiotics and medicinal antibiotics should be separately identified. The medicinal antibiotics used by vets could remain the same as those used in clinical medicine; the assumption was made that vets, like their clinical colleagues, are responsible practitioners and would prescribe antibiotics carefully. On the other hand, antibiotics used for growth promotion would now have to be different and unrelated to those antibiotics used for medicinal reason. The difference is sometimes difficult to quantify. These regulations were not imposed elsewhere in the world but were eventually adopted in Europe. In the United States, however, the use of medicinal antibiotics was continued for growth promotion. By the late 1980s, resistance rates to tetracycline in *Escherichia coli*, isolated from food-producing animals in the United States, was 96%. Resistance to ampicillin was only slightly lower at 77% and although tetracycline usage may be in considerable decline, partly as a result of increased resistance, ampicillin and the closely related amoxycillin are the most widely used antibiotics in the world.

The problem could be further demonstrated with pigs which had been fed a tetracycline derivative prophylactically. During the treatment, their faeces were full of tetracycline-resistant bacteria but after withdrawal of the antibiotic, these resistant bacteria seemed to disappear. The introduction of short courses of tetracycline revealed that the tetracycline-resistant bacteria had not been eradicated but small numbers had survived in the gastrointestinal tract which would have preferential survival in the subsequent use of the antibiotic. Withdrawal of the antibiotic did not remove the resistant bacteria, only suppressed them. When the pigs were treated again with tetracycline, resistant bacteria soon predominated. The speed at which these bacteria emerged could not be explained by the development of resistance; resistant bacteria must have been established in the gastrointestinal tract and their presence masked in an apparently sensitive population.

It is easy to demonstrate that the use of antibiotics in animal husbandry leads to resistance; animals reared intensively carry bacteria with much higher incidences of resistance. In a comparison study of resistance carriage in *E. coli* in range-reared cattle and those reared intensively, there was no tetracycline resistance and only 1% resistance to ampicillin. On the other hand, in the intensively reared animals, 50% of the *E. coli* isolated were

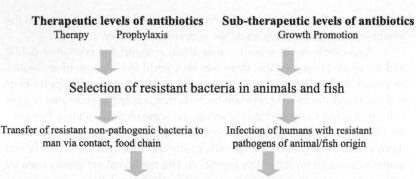

Therapeutic levels of antibiotics   Sub-therapeutic levels of antibiotics
Therapy   Prophylaxis   Growth Promotion

Selection of resistant bacteria in animals and fish

Transfer of resistant non-pathogenic bacteria to   Infection of humans with resistant
man via contact, food chain   pathogens of animal/fish origin

**Acquisition of "new" resistance genes by human
commensal and pathogenic bacteria**

Consequences of antibiotic usage in animal husbandry and aquaculture

tetracycline resistant and 13% were ampicillin-resistant. Indeed, considerable differences in resistance carriage rates can been seen within the United Kingdom; in Scotland where there is more agricultural land per head of the human population, the cattle tend to be range-reared and 15 years ago, when we examined resistance levels to trimethoprim, they were found to be around 1% in *E. coli* isolated from cattle. In England, where cattle are more intensively reared because agricultural land is at a greater premium, the resistance rates were found to be very much higher. Usually the less intensively farmed, the less likelihood of disease and a corresponding transmission of resistance from one animal to another.

The role of growth promoters in the selection of resistance
If bacteria are constantly exposed to growth promoters then surely all should eventually become resistant, but not all resident bacteria are resistant. The strongest argument why this has not occurred is that the concentration of growth promoters is so low that they are simply in insufficient concentrations to cause resistance. The argument continues that for resistant bacteria to predominate, they must be positively selected and this can only occur if the sensitive bacteria are inhibited at concentrations around or above the minimum inhibitory concentration. This argument seems false for two reasons; the first is that we have already seen that this is unnecessary and sub-inhibitory concentrations are more than adequate to promote resistant bacteria. The second is that growth promoters are thought to be beneficial *because* they inhibit gastrointestinal bacteria. Under these conditions, the emergence of resistance would be thought to be a natural consequence of the use of growth promoters.

In many farming areas in Denmark, a bacterium common to the faeces of most terrestrial farm animals, *Enterococcus faecium*, was rapidly becoming resistant to vancomycin, an antibiotic used in clinical medicine but never used in veterinary practice. To explain these observations, it was thought that a closely related antibiotic was being used as a growth promoter and that its use was selecting resistance to both antibiotics. The most likely candidate was the growth promoter avaparcin which, like vancomycin, is a glycopeptide antibiotic. This growth promoter has been available in Europe for over 20 years and has been used to boost yields in most livestock. It can be given in very low concentrations, starting at around 5 parts per million, usually rising to a maximum of 40. Even at the low concentration, it markedly improves growth rate and conversion of feed into muscle. As this drug is more effective against gram-positive bacteria, presumably it boosted yields by inhibiting the gram-positive bacteria of the gut. The emergence of resistant *Enterococccus faecium* would be an inevitable outcome of avarparcin use.

The dilemma is whether the avaparcin resistance gene is responsible for vancomycin resistance. If it is, then vancomycin resistance is being selected directly by the use of a growth promoter in animal husbandry and the use of avarparcin would become unacceptable, particularly bearing in mind the predominance that vancomycin has in the treatment of serious gram-positive infections in hospital, particularly methicillin-resistant *Staphylococcus aureus*.

When Sweden joined the European Union in 1995, it had already banned the use of growth-promoting antibiotics for nine years. The Commission gave Sweden three years to drop the ban unless it could convince the other states that the ban should remain and be imposed throughout the EU. Denmark was keen to introduce a ban because of an alarming increase in resistance and Germany followed, so both countries had a ban imposed by early 1996. These bans were challenged by the EU Scientific Committee on Animal Nutrition (SCAN) in July 1996 who examined the scientific evidence submitted by the countries that had imposed a ban and then rejected it on the basis that a ban of a growth promoter sets up an unfair trade barrier. The European Commission ignored SCAN's recommendation and proposed a ban of avaparcin use throughout the EU in December 1996. This ban was based on "scientific opinion" though details have not been provided. The British followed with a ban in April 1997.

Whatever evidence existed was scanty and circumstantial. Enterococci are part of the normal faecal bacteria of mammals and poultry, so animals treated with avaparcin could be considered to be a significant threat to the community as they might act as a vast reservoir. This reservoir would continuously be challenging the human population and there are examples of vancomycin-resistant enterococci entering the food chain. Bates and colleagues have demonstrated vancomycin-resistant enterococci in frozen chickens bought in a supermarket. Even more alarmingly, it was shown that chickens delivered to hospital kitchens also contained vancomycin-resistant entero-cocci. These bacteria have also been reported in minced or ground beef.

The most sophisticated epidemiological studies, which have examined the DNA of the enterococci isolated from raw sewerage and livestock and then compared them with the enterococci isolated from humans, have shown that all the bacteria are identical. This strongly suggests that bacteria are migrating from one environment to another, probably from farm animals to man. Recently, resistant enterococci have been isolated from a community patient and then compared with enterococci isolated from porcine faeces. The patient had neither been in hospital nor had taken any antibiotics. This led to the assumption that the patient had ingested pork infected with resis-tant enterococci. It had always been assumed that resistant enterococci were selected and emerged in hospitals under excessive vancomycin usage; however, this hapless patient led to the theory that vancomycin resistance was actually selected in the community by the unwitting ingestion of resis-tant bacteria. This evidence is circumstantial and the numbers are too low to show any direct link with the acquisition of vancomycin resistance from infected meat. Even if a direct link could be shown, it does not necessarily mean that avaparcin is to blame.

In principle, many microbiologists would welcome the recent ban on the use of avaparcin, which will prevent the widespread use of an antibacter-ial substance particularly used as a growth promoter, but the evidence for the ban on avaparcin seems scanty. There is widespread vancomycin resistance in the faeces of pets in Europe and, as far as is known, neither avaparcin or vancomycin is used in cats or dogs. It could be argued that the resistant bac-teria have entered the food chain or the residues have been passed on in pet food and are selecting resistant bacteria in the pets themselves. The most compelling evidence that avarparcin may not be the culprit is that there is nearly as much resistance to vancomycin per head in cattle in the United States as there is in Europe; the United States has never allowed the use of avarparcin as a growth promoter or even as an antibiotic in animals. The ban seems political; the correct view nowadays is that we should use as few chemicals in agriculture as we can and growth promoters and preservatives are high on the unpopularity list. What would happen if we were to remove them from food production? Certainly food would cost much more but it would likely be more contaminated than it currently is, even though the con-taminating bacteria might not be resistant.

Intensive agriculture is not the only change that has taken place in the twentieth century; we now all have access to refrigerators. In these refrigera-tors we keep both cooked meat and uncooked meat. We have a preference towards frozen poultry and minced or ground beef. *Salmonella* are regularly found in the carcasses of intensively farmed chickens. These bacteria are now found in approximately 30% of all frozen intensively farmed birds and we accept this as part of modern-day living. Chickens represent a lesser financial investment than cattle or pigs and are treated less regularly. Therefore bacteria infecting these animals have been exposed to fewer antibiotics and antibiotic resistance is concomitantly lower although frequent outbreaks demonstrate the potential threat that these animals pose. Minced beef also poses a significant threat; muscle is impervious to bacteria so the organisms would not normally get inside a piece of steak. The cooking of such a slab of meat would readily kill the microorganisms on the surface. In fact, the heating could be quite modest to kill all the surface bacteria. This why a rare steak can be virtually free of all bacteria. Once beef is minced all the meat becomes contaminated with the bacteria that had previously been confined to the surface. Unless rigid washing and decontamination of the mincing machine is carried out between each meat batch, cross-contamination is likely to be common. Decontamination between meat batches is, for most purposes, impractical so contamination is frequent. The action of mincing meat takes the outside edges, contaminated with bacteria, and grinds them into the middle of the meat mass. Thus, unlike steak, minced beef has contaminating bacteria in the centre. The minced beef is then often used for hamburger production and, unless the meat is extremely harshly cooked, the middle will not reach a sufficient temperature to kill any residing bacteria. This is considered to be a major contributory factor to the infection of *Escherichia coli* O157, which has caused so many problems throughout the world but particularly in Scotland.

In essence, it should not matter to us whether our meat is contaminated by bacteria on the outside or whether these contaminating bacteria are resistant to antibiotics. The hygiene in our kitchens and our cooking practices should prevent any infection by these bacteria. The reality is sadly often different. In commercial meat preparation cooked and uncooked meat should be kept in different cabinets; how many domestic households have a separate refrigerator to keep uncooked meat separate from all other food? How often are work surfaces washed down and disinfected after uncooked meat has been prepared? Our preference is now for minced beef, mainly because it is cheaper, and we often eat a slab of minced beef as though it was a piece of steak. Many like this slab of mince exposed to the minimum of heating to produce a rare hamburger. The risks of eating a slab of mince cooked to such a limited extent are great. My PhD supervisor was so insistent that mince prepared for him was sterilised when cooked that he was always sending it back in restaurants for further cooking until it was once written on the bill, "a hamburger cooked until the chef cried". He never suffered from food poisoning because his meat was always sterile. We complain about contamination of meat from bacteria, sensitive or resistant to antibiotics, but the solution is with the consumer and the commercial food preparer to improve domestic and commercial hygiene.

Therapeutic use of antibiotics in animals
The use of antibiotics in the therapeutic treatment of animals seems a much greater threat. The quantities of drugs used are very much greater than for growth but, more importantly, the drugs that are used are almost identical to those employed in human clinical practice. Unlike human clinical practice, the animals are treated on a herd basis, thus if one animal is infected, the remaining herd is treated prophylactically. Apart from the fact that cattle and pigs might have a greater body mass than humans so more antibiotic will be used per animal, the amount of antibiotic challenging the herd is the amount per animal multiplied by the number in the herd. This is often a very large figure and makes the quantity used in growth promotion quite paltry. This massive antibiotic usage selects resistant bacteria. Apart from vancomycin-resistant enterococci, the resistant bacteria most implicated in human infection are *Campylobacter jejunii, Yersinia enterocilitica* and *Escherichia coli*. Each of these bacteria can cause gastrointestinal infections in man but the evidence of human zoonotic infection (infection caused by animal bacteria) is circumstantial. The only clear demonstration of human infection by resistant bacteria is gastrointestinal infection by *Salmonella* spp. Outbreaks of gastrointestinal infection caused by *Salmonella* are often reported and these have increased considerably since the advent of intensive farming practices and the freezing of poultry carcasses.

In a study about 12 years ago, it was demonstrated that a number of American patients became seriously ill with gastrointestinal infection within 1–2 days of starting antibiotic therapy. These patients had had no history of gastrointestinal infection and had been treated with antibiotics for infections at completely different sites. The rapid emergence of gastro-enteritis suggested that resistant bacteria must have already been present in the gut and the use of antibiotics merely provided the selective environment for these bacteria to proliferate. One of the infections was so rapid and severe that the patient died. The bacterium implicated was *Salmonella newport*, which was residing in the gut in insufficient numbers to cause infection but which was multi-resistant. The bacteria contained a 38-kilobase plasmid which carried a number of resistance genes. Beef cattle used as the food source for these patients also carried *Salmonella newport* which also carried the same 38-kilobase plasmid. These beef cattle were fed chlorotetracycline in subtherapeutic concentrations for growth promotion and chlorotetracycline resistance is one of the genes carried by the 38-kilobase plasmid. I saw a similar incident when I was working on my PhD and one of my colleagues was working on *Salmonella typhimurium* containing the plasmid R46. Many microbiologists can harbour some of the bacteria that they are working on, without manifesting any clinical symptoms. When this PhD student had a gastrointestinal infection and went to the University medical centre she was prescribed ampicillin. Plasmid R46 carries a resistance gene for ampicillin. She suddenly had severe gastro-enteritis; the antibiotic had suppressed all the sensitive *Escherichia coli* in the gut and provided the environment for *the Salmonella typhimurium* to proliferate. The student became very ill and had to be treated for severe salmonelleosis. The point is that resistant bacteria can reside in small numbers and emerge only when antibiotics are administered, often for infections that might be in quite unrelated areas.

There has been much concern that the veterinary antibiotic apramycin has selected out resistance genes that also confer resistance to aminoglycoside antibiotics used in clinical medicine. A diagram for the potential mode of transmission of these resistance genes is shown in the figure below. However, evidence for direct transmission of resistant bacteria from animals to man is scanty. Chloramphenicol-resistant *Yersinia enterocolitica* obtained from man has been shown to be identical, as far as current tests can demonstrate, to the same organism isolated from pigs. This could lead to the speculation that the porcine bacteria are, after ingestion, responsible for human infection. In 1984, in Edinburgh, there had been an epidemic of a plasmid, called pUK28, in clinical *Escherichia coli* isolated from patients both in hospital and in the community. This plasmid conferred resistance to ampicillin, streptomycin, sulphonamides and trimethoprim, the main antibiotics used for the treatment of common infections at that time. The source of this 74-kilobase plasmid was a mystery and the faeces of pigs entering the markets of Edinburgh were examined for the carriage of trimethoprim-resistant bacteria. In Scotland, sheep and cattle are kept in the open so are rarely treated with antibiotics and the incidence of animals carrying trimethoprim-resistant *Escherichia coli* was less than 2%; none of them carried plasmid pUK28.

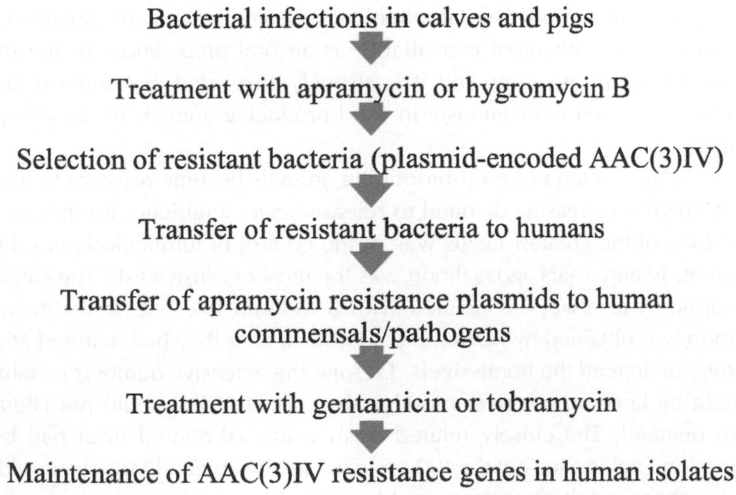

Bacterial infections in calves and pigs

⬇

Treatment with apramycin or hygromycin B

⬇

Selection of resistant bacteria (plasmid-encoded AAC(3)IV)

⬇

Transfer of resistant bacteria to humans

⬇

Transfer of apramycin resistance plasmids to human commensals/pathogens

⬇

Treatment with gentamicin or tobramycin

⬇

Maintenance of AAC(3)IV resistance genes in human isolates

Transfer of Apramycin resistance from "animal" to "human" bacteria

Similarly, there was no trimethoprim resistance in the faeces of chickens despite the fact that they are intensively farmed mainly in units to the west of the city. Only pigs are intensively reared and the incidence of pigs with trimethoprim-resistant bacteria was around 18%, much higher than the other animals. These pigs largely came from farms clustered around the south-west of the city and could be implicated as the source of these resistance genes. However, detailed examination of the plasmids revealed only one pig in the whole study carrying an *Escherichia coli* harbouring plasmid pUK28. Bearing in mind that this plasmid accounted for 33% of trimethoprim-resistant bacteria in hospitals and 25% in the community, its rarity in porcine bacteria did strongly suggest that pork was not the source. Indeed, it could be argued that human bacteria had infected the one pig in which plasmid pUK28 had been found.

An interesting observation was the use of trimethoprim in food-producing animals. During the 1980s there was increasing concern that the combination of trimethoprim with a sulphonamide was increasingly difficult to justify because of the adverse effects of the sulphonamide. However, the same manufacturers who promoted trimethoprim alone in human medicine still marketed the combination for use in animals. Despite the publicity related to the perceived risks associated with the combination, no such concern was expressed about the animal preparation. It could be argued that the use of the combination in animals could result in sulphonamide residues finding their way into man.

Direct implication may be very difficult to prove but it may be that antibiotics are being used in cases for which they were never intended. In the United States, the use of chloramphenicol is now restricted to non-food-producing animals but still chloramphenicol-resistant bacteria are isolated from animals; chloramphenicol is available as an oral preparation for the treatment of canine infections and it is strongly suspected that some of these products are used intravenously in food-producing animals for growth promotion.

As the bacteria of food-producing animals become resistant to antibiotics, there is increasing demand to release newer antibiotics for therapeutic use. One of the greatest needs was in the control of furunculosis in salmon farming. Twenty years ago, salmon was the most expensive fish. The catching of salmon was always associated with a rod and line but, in reality, most salmon was obtained by netting the adult fish, after they had matured at sea, as they re-entered the home rivers. Despite the extensive numbers of salmon caught by large nets at river mouths, the supply of fish could not keep up with demand. The closely related fresh water salmonoid trout had been farmed in England for nearly 400 years so it was argued that pens could be set up at sea in which salmon could mature in conditions which simulated their development once they had migrated to sea.

Farms were set up in the sea lochs of the west coast of Scotland which comprised a series of netted pens that were suspended from the surface. The bottom of these pens would often be only a few metres above the sea bed. In each of the pens large numbers of salmon smolts are released and fed high-energy pellets. As the fish increase their body weight, the pens become crowded; each pen might contain hundreds of thousands of fish. Under these conditions, the fish become stressed and prone to disease. In Atlantic salmon, the infection they become most prone to is furunculosis. This is a kidney infection cause by the bacterium *Aeromonas salmonicida* var *salmonicida*. The name of the disease comes from the liquifactive muscle lesions (furuncules) that are formed on the back of the fish. Interestingly this is not the most severe manifestation of the infection and the fish die more readily when no furuncules form and septicaemia develops. Furunculosis can occur in wild salmon but outbreaks are rare; it is much more common in farmed fish. The transmission of the bacterium is through water from a contaminated fish.

Once salmon farming became established, the number of farms reporting problems with furunculosis escalated, rising from 10 to 1979 to 137 in 1989. This was accompanied by a severe reduction in the percentage of mature fish recovered from the smolts entered into the pens. In 1984, there was 86.5% recovery whereas, by 1987, this figure had dropped to 65.5% while the number of smolts entered into sea pens had increased fourfold to nearly 13 million. The increasing number of smolts entered into the pens was leading to increasing mortality. Without the introduction of therapy to control the furunculosis, the whole salmon farming industry was threatened. The problem could be alleviated by a considerable reduction of the number of smolts in each pen but this would raise the costs of production and boost the price of the fish in a highly competitive market.

The industry turned to antibiotics but no drugs had been specifically licensed for fish. Indeed no specific trial had ever been performed on salmon. The only antibiotics available were those that were used against cattle, sheep, pigs and poultry and veterinarians had to use these antibiotics to treat furunculosis. The problem of administration of these antibiotics was that it was impossible to dose individual fish and, in any case, it would be entirely uneconomic in pens that contained in excess of 100,000 fish. The only method of administration was to add antibiotic to the food pellets. This caused two problems; the first was that antibiotics are relatively unpalatable and whereas we might be persuaded to take them as we are aware of the benefit, fish have no such incentive and will not touch food they consider distasteful. The second problem is that the fish lose their appetites after contracting furunculosis and will not eat the antibiotic-coated pellets. Therefore antibiotic administration can only be given before the animal develops symptoms, so effectively it is administered prophylactically.

In 1983, the early quinolone oxolinic acid was officially licensed for the salmon farming industry. This was followed two years later by the licensing of oxytetracycline and then in 1987, the combination of trimethoprim and a sulphonamide was approved as Tribrissen. The intensive usage of these antibiotics quickly led to resistance. By 1990, in Scotland there was widespread resistance to oxolinic acid and oxytetracycline amongst *Aeromonas salmonicida* isolated from infected fish in farms all over the country. The oxytetracycline resistance was carried on plasmids so could be transferable from one strain to another and this might explain its rapid spread through the fish pathogens. The resistance to oxolinic acid was much more problematic. This drug is poorly absorbed and might reach the site of infection in low concentrations, and it is only able to kill bacteria slowly; however, its greatest disadvantage is that resistance emerges far more quickly to this drug than to most others. These factors combined have ensured that the large numbers of viable bacteria that remain within the fish during treatment are readily able to mutate to resistance. This augurs badly for further treatment because acquisition of resistance to oxolinic acid usually results in partial cross-resistance to the fluoroquinolones. Thus if further treatment is eventually required with more powerful drugs, their success may have already been compromised. This is an example where a weak drug was being used because it had been felt that the market did not justify more powerful members of this class of drug. In fact, this is a false assumption; the weaker drug does not work efficiently and, more importantly, this very weakness leads to resistance which severely restricts the potency of more powerful drugs. Modern perception of antibiotic usage suggests that the most powerful drugs within a class should be used early if cross-resistance is likely to be a problem.

In September 1990 amoxycillin was licensed for use for the treatment of furunculosis. It was a strange choice for the treatment of a disease caused by *Aeromonas salmonicida* because this genus is known to produce β-lactamases capable of destroying amoxycillin; however, amoxycillin had already been licensed for veterinary use and so the progression to licensing for use in aquaculture might be smoother than it might have been for other drugs. It was also relatively cheap and, in the quantities that it would be required, cost was a major consideration. In reality the need to control furuncluosis caused by *Aeromonas salmonicida* was becoming so great that almost any new antibiotic would have been considered. Once the antibiotic was introduced, resistance did emerge and strains were examined for their resistance mechanisms. As expected, they did contain a β-lactamase that conferred resistance to amoxycillin; however, they also contained another two β-lactamases, one that conferred resistance to cephalosporins and one that gave carbapenem resistance. This meant that the use of amoxycillin was selecting bacteria that were not only resistant to penicillins, the antibiotics being used, but also to two groups of β-lactams that had never been used against this organism. Cephalosporin resistance is quite widespread but carbapenem resistance is not and this class of β-lactam is considered to be the final defence against many hospital infections. The selection of resistance genes to this drug in farmed fish was considered to be an undesirable feature as it is conceivable that these genes could become incorporated into clinical bacteria in the future.

The use of antibiotics in fish farming became more unpopular and less acceptable. It was found in Scotland that most of the antibiotic was not eaten by the fish but actually fell to the bed of the sea loch. The shape of the Scottish sea loch is usually a gently sloping bed so that the residues usually remain underneath the pens. Examination of the beds under the pens often revealed destruction of the local flora. This raised many concerns amongst environmentalists. In Norway, fish farming is at least as active as in Scotland and there is far less environmental concern. The farms used antibiotics as much as their Scottish counterparts; but the shape of a Norwegian fjord is very different from that of a sea loch; fjords are very steep-sided and the pens are often located over much deeper water. It is often impossible to study the area under the pen as the water is too deep. This meant that the effect of the antibiotics was merely less visible.

The easiest solution to control furunculosis is to reduce the concentration of fish in the pens. The need for antibiotics is reduced considerably when the concentration is lowered. In the early 1990s, heavy financial competition occurred as the salmon market in Europe was flooded particularly by fish reared outside the European Union. The price of fish fell to around £4 per kilogram which meant that reductions of the fish concentration would render most ventures uneconomic. The rise in amoxycillin resistance meant that further antibiotics would have to be introduced with the inevitable development of resistance to them and the further pressure that they might put on the immediate environment of the pens. There was much discussion that the fluorquinolone enrofloxacin should be used. This was a powerful drug that could overcome resistance that had already developed to other members of this drug class and resistance would be less likely to develop. The dilemma was resolved not by an antibiotic but by a vaccine and almost all salmon are now vaccinated against *Aeromonas salmonicida* infections. This has proved far more successful than antibiotic treatment and has no direct environmental impact.

ENROFLOXACIN

Older quinolones were used in veterinary medicine for a decade or more before fluoroquinolones were introduced into clinical medicine. When ciprofloxacin was demonstrated to have remarkable capabilities to control human infection, the benefits for animal husbandry looked attractive. A number of new fluoroquinolone derivatives were made, many specifically for veterinary use, These included drugs such as benofloxacin, ofloxacin, danofloxacin and enrofloxacin. The latter drug enjoyed the most popularity and was a direct derivative of ciprofloxacin. It proved particularly useful for the treatment of diarrhoeal diseases in chickens. None of these drugs were licensed by the Food and Drug Administration for use in food-producing animals although they could be used for the treatment of infections in domestic pets. The reluctance to license the drugs for agricultural animals rose from a fear that the use of these drugs would shorten their therapeutic lives in clinical medicine because their widespread use would promote the selection of resistance. They were introduced into veterinary practice in Europe and these reservations of the Americans now seem justified.

Ciprofloxacin resistance amongst *Campylobacter* spp isolated in poultry products and humans in the Netherlands is increasing. Between 1982 and 1989, the prevalence of resistance amongst isolates from chickens increased from 0 to 14% while the corresponding figure in humans rose from 0 to 11%. The matching rises in the figures for fluoroquinolone resistance were taken as an indication that the problem in human bacteria was a direct result of veterinary administration. Certainly, norfloxacin and enrofloxacin had been introduced over that time period, norfloxacin to treat human infection and enrofloxacin in veterinary practice. Campylobacter infection in humans is usually considered to be the direct result of infection from ingested poultry or from working with uncooked poultry products. The hypothesis rests on the assumption that the bacteria both in humans and poultry have never been exposed to quinolones before. This is, of course, an invalid conclusion because both sets of bacteria had already been exposed to the older quinolones for at least 10 years before the introduction of the fluoroquinolones. This had been nalidixic acid for human bacteria which had been introduced in the mid-1960s and oxolinic acid and, particularly, flumequine to treat animal infections. When it was stated that there was no resistance in 1982, resistance was measured against the modern fluoroquinolones and almost certainly would be undetectable if conventional sensitivity tests and breakpoints were employed. If the bacteria had been tested for sensitivity to the early quinolones, which they were not, then the result would have likely been very different; it would have been probable that the bacterial sensitivities to these drugs were already starting to decline. This point is important because the very rapid increase in resistance to the fluoquinolones in the 1980s is surprising; resistance to the fluoroquinolones does not usually develop as rapidly because, to attain clinical significance, it usually has to be made up of a series of mutations, maybe as many as three, and time is often required for these mutations to establish sufficiently for the next mutation to occur.

It is much more likely that the bacteria that were later shown to be resistant to clinical levels of fluoroquinolones already had a "head start" and were resistant to older quinolones, so the initial mutations required for fluoroquinolone resistance had already taken place. If this was the case, then the response of bacteria to the introduction of fluoroquinolones would probably just be a single mutation which would build on those that had already taken place. This would explain the very rapid increase found once fluoroqinolones were released. The difficulty comes in trying to prove this hypothesis, especially so long after the event. In order to demonstrate direct transmission of fluoroquinolone-resistant bacteria from animals to man, extremely sophisticated typing techniques would be required and they were not available at the time. It would have to be shown that the bacteria isolated from man and food-producing animals were identical in every way and, equally important, that the exact resistance mechanisms were identical as well. This is only just achievable now and at that time it was assumed that almost all resistance was a result of alteration of the *gyrA* gene inducing changes in the structure of DNA gyrase. We now know that it is much more complicated than that and that other targets for the fluoroquinolones exist; alterations in these can cause resistance, as can changes in permeability of the antibiotics.

We might have the opportunity to assess whether the introduction of enrofloxacin in poultry production affects resistance in animals in the United Kingdom. The antibiotic obtained a licence in 1994 so if there is a corresponding increase it should soon be apparent. Again the first problem in assessing the influence of this type of introduction is the previous quinolone usage. It has been shown that *Salmonella* spp isolated from animal specimens have demonstrated increasing resistance to the early quinolone nalidixic acid in the six years prior to the introduction of enrofloxacin. It has also been shown that this also decreases the susceptibility to the fluoroquinolones; however, it does not necessarily result in demonstrable resistance to the fluoroquinolones, as the level of insusceptibility to ciprofloxacin is still below the breakpoint level set to indicate clinical resistance. Even in 1997, the introduction of enrofloxacin does not appear to have increased resistance to fluoroquinolones in man. Interestingly, it does not seem to have increased resistance to enrofloxacin in British poultry bacteria to the same extent as had occurred in bacteria isolated in European poultry. This might result from the fact that there was more resistance to the older quinolones before the introduction of fluoroquinolones and supports the view that the older drugs are the most powerful selecting agents. Essentially the experiences in the United Kingdom do not resolve the debate as to whether resistant animal bacteria are passed on to Man. In fact, it is probably more complicated in Britain, because the United Kingdom imports about 20% of all the meat consumed.

TO LEGISLATE OR NOT TO LEGISLATE

In order to demonstrate that resistant bacteria from animals are the progenitors of resistant bacteria in man, the very least that has to be demonstrated is
that the bacteria are identical by all genetic typing techniques and that not
only are the resistance mechanisms identical but they are manifested by
exactly the same mutations. These criteria may never be achieved in the
comparison of resistant bacteria in animals and man. In the absence of these
epidemiological data, no conclusion can be made that resistant animal bacteria are responsible for infecting man. It should be emphasised that the
information listed above is the *very least* that is required and we may find in
the next 10 years that even more powerful discriminatory techniques can
demonstrate that bacteria which we previously considered to be identical
are not. Biological scientific research is essentially the ability to demonstrate
differences; all experiments should be compared against controls and the
experiments identify variations from the controls. It is virtually impossible to
state that two parameters are identical and this is what is required when we
seek to demonstrate that bacteria from animals are infecting man.

It could be argued that this is just being pedantic and that all that is
required is a demonstration that the use of antibiotics in food-producing
animals and the development of resistance in their bacteria are likely to
cause to resistance in man. If this presumption can be made, this is sufficient
to restrict the use of those antibiotics in animal husbandry. This is essentially
how the decisions are currently reached and the withdrawal of antibiotics
from veterinary use is based on trends rather than hard molecular epidemiological fact. It is essential that we are able to make decisions based on fact
rather than assumption. There are, of course, many who would like to see the
use of antibiotics removed from all forms of animal care. The removal of
antibiotics would force a reduction in the intensity of farming in this country
and in many in the developed world. In the United Kingdom and the United
States, a small reduction in intensity could be sustained, as farming has
become so efficient that land is deliberately being set aside because it is no
longer required; but the removal of antibiotics would limit the capability of
England to provide sufficient animal carcasses to feed its population. There
would be a marked increase in food prices, and the lack of take-up by the
majority of the population of organically grown food, which is significantly
more expensive, suggests that the increased cost of food produced in the
absence of antibiotics would not be acceptable to the average consumer.

It is, of course, essential for public health that an antibiotic is withdrawn immediately if its use is clearly demonstrated to select resistance in
animal bacteria that was passed on to man. I would not wish to see regulations that allowed any lesser action; however, the argument is not convincing
if the evidence is anecdotal or if the data presented are incomplete. The economic implications of unnecessary withdrawal are massive and so should be

made only on the most rigorous data. Unfortunately, governments often do not take this into account and most investigations are performed by those scientists who have an academic interest in the subject or by the pharmaceutical companies which might have a financial interest in the outcome. There is virtually no control over the data that are generated and no guidelines on the validity of the data needed to legislate for withdrawal. This is an important matter for the industrialised countries to consider; if their own meat production is to continue to provide their own needs, then they must provide firm guidelines on the information required for decisions to be made on the future use of antibiotics.

CONCLUSIONS

Ever since the Swann Report, it has been assumed that resistance in human bacteria has been a direct result of antibiotic usage in animals. I think that in the intervening 31 years, it largely remains an unresolved problem but still raises much passion and the veterinary usage is regularly blamed for the increases in resistance seen in clinical bacteria, though often little mention is made of the capability of clinical usage of antibiotics to select resistance itself. We tend to see this enigma very much in our own environment. We shall see in the next chapter that by far the greatest resistance problems exist in the developing world and in the Indian subcontinent. India is an agricultural country and could feed its population, which is rapidly approaching a billion. Antibiotics are relatively expensive and too expensive to be used by farmers, who tend to be poor, for the treatment of infections in their animals. Resistance levels in bacteria isolated from animals in India tend to be low by comparison with resistance in the intensively reared animals in the industrial countries. Often in rural communities, the animals live in close proximity with the human population, frequently in the rainy season under the same roof. In these communities, antibiotic resistance is extremely high; however, no contribution to this can be attributed to either antibiotic usage in animals or the transmission of resistant bacteria from animals to man. Antibiotic usage in food-producing animals is not a prerequisite to resistance in humans and it still remains unclear how significant its contribution might really be.

8

The doomsday scenario – antibiotic resistance in the developing world

It is often surprising to the first-time visitor to the Third World that in the middle of horrible deprivation, there are often well-stocked pharmacies with products that would be found in pharmacies in industrialised countries. Even in economies where the cost of living is much lower than it is in Europe, the cost of these antibiotics is equivalent to that in the industrialised countries; therefore, antibiotics can represent a very much higher financial burden than they do for us. A single tablet of a fluoroquinolone can cost more than a day's wages for an agricultural worker in India, the equivalent of perhaps £25 per tablet in British terms or $US75 for an American. In a free market, the cost of a product or commodity is determined by demand and the cost of tablets in relation to a day's wages represents the importance, almost magical capabilities, attributed to these drugs. It is doubtful that if tablets cost £25 or $US75 each, whether antibiotics would retain their prominence.

When I was on a field trip to India in 1989, we decided to find out how antibiotics were distributed and what was the local perception of their usage. I went into pharmacies in South India and explained that I was taking a trip up-country and that I was worried about contracting a gastrointestinal infection or even typhoid. I asked each pharmacist if he could recommend and sell me a full course of antibiotics that he considered suitable to do the job of controlling such infections. Those who considered me to be an affluent foreigner suggested fluoroquinolones which included some drugs that were not yet available in Europe. Others to whom I pleaded that, although I wanted a full course of drugs I was not affluent and that cost would be an important consideration, tended to steer me towards courses of amoxycillin, co-trimoxazole and tetracycline. There was no fixed price for any of these antibiotics so I was able to bargain to lower the price for all the antibiotics that I bought though I am not nearly as skilled at this practice as a local resident.

The important points of these encounters were that I was able to purchase the antibiotics over the counter, a practice that will be discussed later in this chapter. The second is that the owner of the pharmacy was recommending the therapy rather than a qualified medical practitioner; it is very unlikely that any of these owners was a qualified pharmacist. The final concern is that the courses recommended *never* matched a full course as accepted in the industrialised world. The owners of the pharmacies were breaking up full courses and then selling as many tablets as they considered the purchaser could afford. In my case, this equation usually meant four tablets, which at that time was the equivalent of one day's treatment in a five-day course. In the case of an agricultural worker this would mean a single tablet. In many cases the recommended courses (usually 20 tablets) were included in the literature associated with the tablets; however, when challenged that the doses were insufficient, most pharmacy owners were adamant that what they had recommended was sufficient to cure all infections, including typhoid. Every "course" of antibiotics sold to me was insufficient to clear even the simplest infections and would be totally ineffective against a serious infection like typhoid.

At that time, the international invasion of Indian commerce was resisted by the government and foreign companies had to use an Indian company to market their products. Many of the large western pharmaceutical companies used this device so many antibiotics were distributed in familiar packaging. Some companies did not want to enter into these marketing devices, including the German company Bayer AG. In 1989, Bayer AG had developed the then most powerful antibiotic, ciprofloxacin, for which they still had approximately a decade's worth of world-wide patent. The reluctance to enter a marketing device meant that ciprofloxacin should not have been available; however, it was readily obtainable in all the pharmacies visited and often heralded as the most potent drug for the purposes for which we had declared we required it. By the end of 1989, as many as 27 pharmaceutical companies in India were marketing the drug. We do not know in how many individual manufacturing units these tablets were made and thus the diversity of active compound. As stated in Chapter 6, we obtained tablets from the Rambaxy Pharmaceutical Company and they matched the profile of Bayer's drug almost exactly. Others have obtained tablets made by some of the other manufacturers and, although some do contain reduced quantities of ciprofloxacin compared to that declared on the label, others contained no ciprofloxacin at all. Indeed some tablets contained no antibiotic at all. We shall discuss which is the greater evil later in this chapter. At the end of 1996, 62 companies were marketing ciprofloxacin in India and in 1998 this figure had risen to 88. There has not been a single product licensed by Bayer and not only had their product been pirated but its long-term efficacy was being compromised by inadequate imitations.

The problem was sometimes further exacerbated by local rural medical practitioners who ran clinics in the villages. The villagers would present often with severe gastrointestinal disease. They would be given a single antibiotic tablet by the doctor and given a steroid injection. The antibiotic would be a feeble attempt to deal with the infection while the steroid would go a little way to reducing the inflammation. The problem with administering drugs such as cortisone is that it acts by immuno-suppression; it depresses the body's immune system to eradicate the infection, which is usually required after antibiotic treatment. The patient would often have to part with a day's wages for this. There would often be no follow-up as the patients simply could not afford to return for further therapy.

Infectious diseases in the developing world are often seen in terms of devastating pathogens, those that cause tuberculosis, leprosy, AIDS, rather than the greater propensity of the community pathogens that we see in the west. It was probably the control of tuberculosis that originally caused the most concern about antibiotic usage in developing countries. The disease is very prevalent in South Asia, particularly in India. It was treated with strepto-mycin, rifampicin or isoniazid. The causative organism, *Mycobacterium tuberculosis*, is a slow-growing pathogen that slowly developed resistance to each new antibiotic that was introduced to combat it. Because the bacterium grows so slowly, the sensitivity tests to demonstrate whether the bacterium is susceptible to antibiotics takes a number of weeks to give a result, by which time the patient should be well advanced with his or her therapy. If the sensi-tivity test demonstrated that the bacterium was resistant then it was often too late to change therapy. Resistance in this organism is, fortunately, always by chromosomal mutation so to prevent resistance a strategy of dual therapy was implemented. This would mean that as soon as the patient was diag-nosed, he or she would be given two antibiotics. The theory was that if chro-mosomal mutation occurred on average at a rate of 1 bacterial cell in 10 million (10^7) then the chances of a single cell possessing mutations that con-ferred resistance to two antibiotics would be 1 in 10^{14}. In the natural state this could never occur; the worst infection might comprise about 10^9 bacter-ial cells. The only flaw with this tactic is that many of the infections treated were already resistant to one of the antibiotics. In this case, when the results of the sensitivity tests were known the antibiotic to which the bacterium was resistant would be withdrawn and the patient would be given a substitute antibiotic, to which the bacterium is now known to be sensitive. The substi-tute is given so that the infecting organism is still treated with two antibiotics to which it is sensitive. This strategy is not very effective and still allows resis-tance to develop because if the organism is resistant to one of the antibiotics, until the result of the sensitivity tests are known, the infection is effectively being treated by a single antibiotic, an environment that promotes resistance in *Mycobacterium tuberculosis*. A revised procedure of triple therapy was implemented which ensured that three drugs were administered at diagnosis and this presumed that the bacterium would be found to be resistant to one of them when the results of the sensitivity tests were known. If the organism was found to be resistant, the doomed antibiotic would be discontinued. If the bacterium was found to be sensitive to all three antibiotics, one of the drugs would still be likely to be withdrawn. The tenet was that the bacterium must always be treated with two active antibiotics to prevent resistance. Triple therapy proved to be a very effective mechanism to halt the progressive increase in resistance in *Mycobacterium tuberculosis* and as long as the patient continued with the therapy, resistance was unlikely to develop. Therapy for tuberculosis often lasts for six months and patient compliance is likely to fade, so progressive health workers in field clinics in developing countries often insist that the patients attend the clinic every day and take their therapy in front of a clinic member. The antibiotics have to be given without payment or compliance becomes erratic. Remarkable results have been obtained, with a complete halt in the development of resistance. Leprosy is caused by a similar slow-growing bacterium, *Mycobacterium leprae*, and triple therapy was introduced to control its spread. The results were remarkable and leprosy could be halted without the development of resistance if administration of the drug was regulated.

RESISTANCE IN COMMON PATHOGENS

When we first investigated antibiotic resistance in the Indian subcontinent in 1984, we did so because we were starting to see changes in the location of resistance genes in common bacteria. The resistance genes were originally located on plasmids and they were transposing onto the bacterial chromosome. This appeared to be the trend in the carriage of resistance genes as it provided a safe and permanent location for the resistance genes in the bacterium, replicating every time the bacterial chromosome replicated and not dependent on the survival characteristics of the plasmid. If this was a trend in resistance development, it could be hypothesised that it would be more advanced in areas where resistance is a greater problem. For many years there had been reports coming from the developing world that resistance levels were higher than they were in Europe or North America. It has to be said that either the studies were poorly controlled or the numbers studied were too low to be significant. The evidence that resistance was high in bacteria from the developing world was anecdotal and entered the folklore of microbiology. *Shigella* spp were often the main cause of gastrointestinal infection and these studies often concentrated on this organism. Although present in the developed world, it is not considered to be a major pathogen and resistance levels are rarely examined in Europe. If resistance in bacteria isolated from the tropics was to be compared with similar bacteria from Europe, then the study had to be performed under identical conditions. Hilary-Kay Young worked in South India and collected common pathogenic bacteria and catalogued them in exactly the same manner as she had with

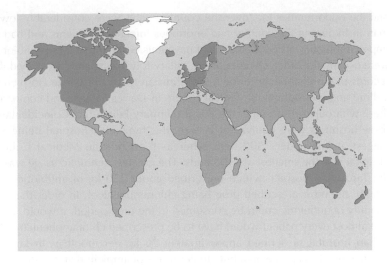

Areas of controlled antibiotic use

similar bacteria from Scotland. She determined the incidence of resistance to amoxycillin and trimethoprim and demonstrated the highest incidences ever recorded. There was 64% trimethoprim and 80% amoxycillin resistance amongst pathogenic *Escherichia coli*. There were even higher incidences of resistance amongst *Salmonella* and *Shigella* spp, the main causes of gastrointestinal infections. The *Salmonella typhimurium* strains carried 12 separate resistance genes, an unprecedented number for the early 1980s, but it showed that antibiotic resistance was out of control and already in epidemic proportions. We soon determined that our original hypothesis that resistance genes migrated to the bacterial chromosome was not true; most of the resistance genes were still located on plasmids but the plasmids were very much larger than they were in similar bacteria in Scotland and some of them carried up to 10 resistance genes. What we were witnessing was the progressive build-up of a resistance armamentarium on plasmids in bacteria that were confronted with many antibiotic challenges. Resistance was still evolving so the stability offered by transposition into the bacterial chromosome was not advantageous while the bacteria still needed these resistance genes in the more mobile environment of plasmids.

Interestingly, there was very little resistance in the bacteria that are usually associated with hospital-acquired infections, such as *Staphylococcus aureus*, in the local teaching hospital of the Christian Medical College Hospital. This was very surprising as resistance problems were already beginning to emerge in equivalent bacteria in Europe. At the same time, we initiated a project in the Muhimbili Medical Centre in Dar es Salaam, the then capital of Tanzania. Tanzania was then under a single-party socialist regime and the opportunities for "over the counter" sales were considerably lower than in India. but the country had enormous infection problems and had to import antibiotics; its poor balance of payments ensured that these antibiotics were usually the older, cheaper drugs. Hilary-Kay Young and I examined the records of the pharmacies of both the Christian Medical College Hospital in Vellore and the Muhimbili Medical Centre in Dar es Salaam and compared these with our own hospital, the Royal Infirmary Edinburgh. Coincidentally, these hospitals were equivalent in size, the Tanzanian hospital being the biggest at just under 1500 beds whereas the Christian Medical College Hospital was the smallest at 1100 beds. The greatest antibiotic usage was in the Tanzanian hospital which prescribed about 250 kg of antibiotics per month, the most prescribed drug being chloramphenicol. In order that this quantity of antibiotic could be consumed in the time-period, it would mean that almost every patient would have to be prescribed chloramphenicol. The Indian hospital prescribed approximately 26 kg per month with far less reliance on chloramphenicol but greater usage of ampicillin and co-trimoxazole. The Scottish hospital used about 13 kg of antibiotics per month with the same spread of drugs that were used in India. The pathogenic *Escherichia coli* were monitored for the incidence of resistance. Not surprisingly, the lowest incidences of resistance were recorded in Scotland with trimethoprim resistance, for example, at approximately 17%. Very surprisingly, the bacteria from the Tanzanian hospital showed only twice the resistance rate at 36%, although 20 times the amount of antibiotic had been prescribed. The Indian hospital showed another twofold increase to 64% although the use of antibiotics was only 10% that in Tanzania. The reason for this massive resistance level was, of course, the influence of the unrestricted use of antibiotics outside the hospital. These comparisons do show that it is often impossible to relate tonnage of antibiotics to resistance levels. The manner in which the antibiotics are used is much more significant than the actual quantity.

The results from Africa seem fairly consistent because our later studies determining resistance levels in pathogenic *Escherichia coli* were performed in the black townships of South Africa and we found similar results to Tanzania. In South Africa the distribution of antibiotics was closely controlled and prescriptions were written for almost all antibiotics. There was some illegal usage of antibiotics but this was not widespread as it was in India.

RESERVOIRS OF RESISTANCE

If there is so much resistance, where does it come from and where are the resistant bacteria or resistance genes when they are not responsible for infections? This largely remained a mystery up until 1989. Some studies had been performed in Europe to examine the carriage of resistant bacteria in the normal gastrointestinal bacteria of healthy people; however, the results were always of special selected populations or the numbers studied were too low

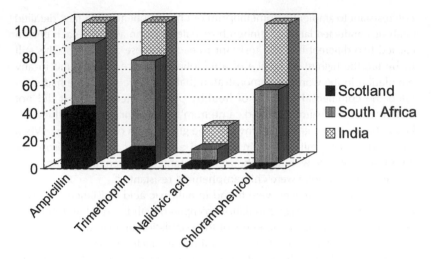

Carriage of antibiotic-resistant strains in the healthy population

to be significant. In any case, the number of resistant bacteria was very low in the populations studied. In 1989, I mounted the only large team field study that I have ever led. The aim was to target groups in South India and obtain faecal samples from them. These samples would be analysed in a laboratory in India and then transported back to Scotland. The group comprised my PhD students, most of whom had never left Europe before and, with staff from the Christian Medical College Hospital, they had to target residents of the town of Vellore and villagers from three villages. The first village, KV Kuppum, was on a main road and had good communications although it was 25 miles from a town. The next village was more remote and located by a river; this village had poorer communications but they were by far the friend-liest. It intrigued them that so many Europeans should be interested in col-lecting their faecal stools. The most remote village, Malmoil, had very limited contact with the outside world and was in the foothills of the Eastern Ghats. These villagers were much more wary of us but were essentially won over by our interpreters. The hypothesis was that the more remote the village, the less exposure the population had had to antibiotics. The latter two villages had no obvious peddler of antibiotics, at least that we could determine.

While we were collected the samples, we asked questions about the domestic arrangements and about the food they had eaten and where they collected water. The laboratory analyses were perhaps the most astonishing that I have ever been associated with. The faecal specimens were placed on agar plates containing ampicillin, trimethoprim, chloramphenicol or nalidixic acid. The first three represented the antibiotics in most common use in the community whereas the nalidixic acid was used as an indicator to detect resistance to quinolones. Almost every person sampled carried *Escherichia coli* resistant to ampicillin, trimethoprim or chloramphenicol. In fact, detailed analysis conducted later in Edinburgh revealed that on average, each person carried two different bacteria resistant to each of these antibiotics. This still is, by far, the highest incidence of antibiotic resistance to be found in any population in the world. It demonstrated that antibiotic resistance had saturated the commensal bacteria of healthy people. Interestingly, it did not matter whether the group studied came from the town or one of the three villages; the resistance genes had invaded the gut bacteria of each group. When we conducted a similar study in Scotland a year later for comparison, we found that only 8% were trimethoprim resistant, 40% were ampicillin resistant and virtually none were chloramphenicol resistant.

The only differences were found in nalidixic acid resistance. This was twice as high in the urban population compared with the rural and the incidence decreased as the remoteness of the population increased.

The reservoirs of resistance for pathogenic bacteria appeared to be the gut bacteria of the healthy population. The resistance genes in the gut organisms were identical to those that had been found in pathogenic bacteria. They were located on plasmids that were also identical or related to those that had been found in the pathogens. It is possible that the resistant bacteria were simply concealed in the gut but we did not, at that time, have the molecular biological techniques to be able to demonstrate that two bacteria were identical. It is more likely that the continuous barrage of low concentrations of antibiotics provides the ideal environment in the gut for the preservation of resistance genes, so when pathogens invade and come into contact with these resistant commensal bacteria, it is easy for the resistance genes to transfer to the pathogens. The water supplies of the region are very variable in

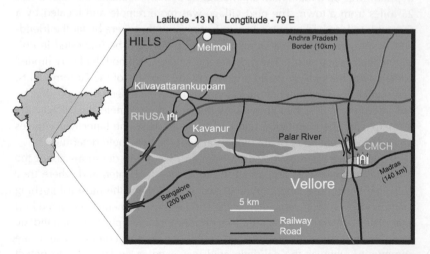

Study villages around Vellore, Tamil Nadu, India

their purity. The deep bore-wells are relatively free of bacteria but the surface water, used by many, is heavily contaminated with resistant bacteria. Resistant bacteria seemed as common in the population who took their water from deep bore-wells; indeed, it did not seem to diminish amongst the few who boiled their water. Thus the source of contamination did not appear to be water. The food sources were very diverse and most was rigorously cooked before consumption. The most poignant social features were the local perception of their own health and the living conditions. Almost 50% of both the urban and rural groups had felt unwell in the previous month. This was usually manifested as a gastrointestinal infection though the number of respiratory infections was also high. In a poor capitalist society, illness means an incapability to work, particularly in the agricultural communities. Therefore most of those who had felt unwell had consulted some medical provider, either at a pharmacy or a local doctor. Almost all had been on some medication which, although it could not always be identified, was invariably antibiotics when it could be. There is no reason to think that this South Indian population was unrepresentative and this points to a massive consumption of antibiotics taken in low doses. The living conditions probably had a significant effect. Most families comprised about six individuals and they all lived and slept in the same room, which in the country was usually located in a hut made of earth. The lack of running water in the dwellings makes hygiene difficult and the cross-infection of bacteria between family members is high, hence the high incidence of perceived infection. The conditions that spread bacteria from person to person are the same that would spread resistant bacteria or even resistance genes. Community bacterial infection is much more common in the poor because of the inability to practise good hygiene and the lack of space; antibiotic resistance is more predominate in bacteria associated with poverty for the same social reasons.

The conditions that promote the carriage of antibiotic resistance became clearer when Dr. Philippa Shanahan, from my laboratory examined the carriage of resistant bacteria in the black towns and villages of South Africa in the period leading up to democracy. The black and white South African communities lived under very different social conditions. The black inhabitants had all the social problems associated with the developing world but with some of the trappings of the developed world. The country had a pure water supply for both communities, and this was at a standard accepted in many developed countries. The distribution of antibiotics was almost exclusively by prescription; the communities were more affluent than those in India and could, if necessary, consult a doctor and pay for any antibiotics. When Philippa examined the commensal bacteria, she demonstrated very high incidences of resistance but not at the level reported in India. She analysed her results according to the age of the study group and found that resistance was higher in the very young; in particular she found that resistant bacteria were readily transferred between children in a day-care centre. Her studies suggested that the social conditions were the main factor in the spread of resistance genes within the population and that the spread of resistance is a manifestation of poverty.

In South America, the incidences of antibiotic resistance are generally high and again many of the communities are poor. No studies have been done to examine the carriage of resistant bacteria in the commensal gastrointestinal bacteria but resistance in pathogenic bacteria is greater than in Northern Europe. The antibiotics are usually made and distributed by reputable pharmaceutical companies but they are generally available "over the counter" in South America. They can be purchased in market stalls in even quite remote parts of Brazil or Chile but, in general, only in the poorest communities is a less than complete course of antibiotics purchased as the vendors are usually reluctant to break up courses of antibiotics. This encourages the patient to take a greater number of tablets and so resistance is generally lower than it is in India, for example. The living conditions are often less deprived and cramped. Resistance in the community is not generally seen as a problem even though it may be as high as 60%; resistance is greater in the hospital and this will be compared in the next chapter.

RESISTANCE PROBLEMS IN DEPRIVED MINORITIES
An example of the contribution that poverty may make to the spread of resistance may be found in the United States. The inner urban minority communities live under conditions more akin to those in a developing country than those of their First-World suburban neighbours. Under these conditions, disease and antibiotic resistance spread rapidly. The most spectacular was the re-emergence of tuberculosis, a disease that is rare under First-World living conditions. It is an associated infection of AIDS and was probably reintroduced into the community by AIDS victims; a severely compromised patient is incapable of eradicating a virulent infection and has to be on almost continuous antibiotic therapy, and this is particularly true if the AIDS patient has tuberculosis. The constant barrage of antibiotics allowed the tubercle bacillus to acquire resistance to each new drug used to attempt to control it. Patients in this urban environment are infected by bacteria resistant to most of the antibiotics, so they often have to commence a therapy regimen that includes four antibiotics in the hope rather than the anticipation that the pathogen will eventually be found to be sensitive to two. It is well documented that AIDS is widely disseminated in sub-Saharan Africa although it is less well known that there is a widespread AIDS epidemic in India. In both areas, many of the sufferers have tuberculosis and other mycobacterial infections, some of which, such as *Mycobacterium avium*, are quite rare in other patients. Treatment of the immuno-deficient patients will

increase the pressure for resistance to start increasing in these bacteria and quadruple therapy may soon become the standard in these areas.

In the poor minority populations of the developed world, it is not just diseases like tuberculosis where problems of resistance manifest themselves. There is ready cross-infection of less severe pathogens such as those of the *Shigella* genus. These bacteria are associated with gastrointestinal infection. They produce a toxin which causes a severe diarrhoea which sometimes contains blood. The most severe pathogen in this genus is *Shigella dysenteriae*, the causative organism for classic dysentery; however, the other species often cause similar symptoms. These bacteria have an extremely low infective dose and it has been estimated that some species require only four viable bacteria to initiate an infection. This is a very much lower dose than all other gastrointestinal bacterial pathogens and it means that only a very few bacteria need to be transmitted from an infected person, thus *any* breach in hygiene leaves the other members of a household susceptible to infection. When I was working in Dar es Salaam, patients in the Muhimbili Medical Centre became infected with *Shigella flexneri*. We could watch the progressive spread of the infection through the wards of the hospital; indeed a diagram of the hospital was used to show the course of the infection and it continued unremittingly until every ward was infected.

About 10 years ago there was a gathering of the so-called "Rainbow Family". This was a group of travelling people comprising New Age travellers and other followers of alternative lifestyles. They met and camped in a National Park in the USA, and there was a massive outbreak of dysentery caused by a *Shigella sonnei*. When the gathering dispersed, the bacterium was spread throughout the United States. The Centre for Disease Control in Atlanta was confirming subsequent outbreaks on the west coast, Hawaii and the eastern seaboard. The bacterium carried plasmids resistant to a series of antibiotics, including trimethoprim, so the dissemination meant that the resistance genes became widely spread as well. Poor personal hygiene and adverse living conditions promoted the spread of this bacterium and the consequential resistance genes.

The spread of *Shigella* species is widespread in South-East Asia and most bacteria have acquired bacterial resistance. The epidemiology of *Shigella* infection in Thailand suggests that infection does not necessarily lead to clinical symptoms; indeed it has been calculated that approximately 90% of those infected do not show symptoms though they do shed large numbers of viable bacteria. It is not clear whether the lack of clinical disease is because the patients have been exposed to previous infections and some immunity has built up. It is clear, however, that the spread of the resistance genes is assisted by the transmission within *Shigella* species, whether or not they cause infection.

Early directives from the World Health Organisation stated that gastrointestinal infections should not be treated with antibiotics; almost all symptoms involving diarrhoea should be treated with oral rehydration which included a large volume of pure water containing sugar and salts to maintain the electrolyte balance and to provide some energy source while the patient was not eating. This is effective therapy for most gastrointestinal infections but it is slow to take effect and the rehydration solutions are unpleasant. The emergence of the newer groups of antibiotics has encouraged some change in policy and antibiotics are often given as well. The cure is effected much more quickly though it is still debated whether a patient treated in this manner sheds more pathogenic bacteria than those treated with oral rehydration alone. In fact, this is likely to be the case, as the early administration of antibiotics allows patients to leave their sickbed and become mobile much earlier; they are still probably carrying the remainder of the infective bacteria. At any rate, the use of antibiotics against infections caused by *Shigella* infections will promote the carriage of resistance genes and the organism will assist in the rapid spread of the genes. Almost all gastrointestinal infections are now treated with antibiotics and oral rehydration appears outmoded. In the International Centre for Diarrhoeal Disease Research in Dhaka, Bangladesh there is now a general policy to include antibiotics in the treatment of *Shigella* infections as well as the treatment of cholera. During the clinical manifestation of cholera, if lost fluids are not replaced, the patient may enter a coma and death would soon follow within 24 hours. Intravenous infusion of saline solution is the optimum treatment but this is very costly and can only be sustained in countries with healthy economies. It was favoured in Chile in the recent cholera epidemic in South America and for the treatment of the few cases that occurred in the United States. However, it is too costly for most countries in South Asia so an inexpensive electrolyte solution is given orally in all except the most moribund. The oral treatment is aided by the concurrent administration of antibiotics.

Vibrio cholerae is a water-borne organism and is a common contaminant in the drinking water of the Indian subcontinent. These pathogens now carry some resistance genes, particularly to antibiotics such as amoxycillin and trimethoprim. This will make them more difficult to treat in the future. These water-borne pathogens are in close proximity with the normal waste faecal bacteria in the rivers and reservoirs. It is very likely that the resistance genes, which are so common in the commensal bacteria, are able to transfer directly to the pathogenic *Vibrio cholerae* rather than resistance developing spontaneously in the cholera bacillus itself. In studies in Africa over a decade ago, it was demonstrated that this organism was already acquiring resistance genes and although some were indigenous to the species, others had clearly been acquired by their contact with commensal gastrointestinal bacteria.

Potential for cross-infection at the International Centre for Diarrhoeal Diseases in Bangladesh

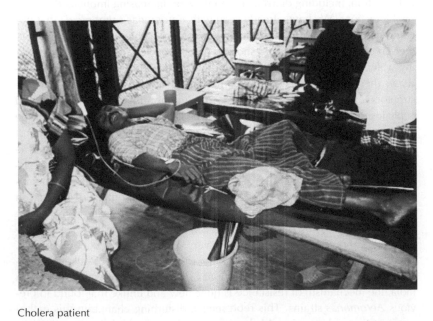

Cholera patient

Cholera epidemics occur sporadically, although every now and again the epidemic escalates and spreads throughout the world to cause a pandemic. This may be associated with a subtle change in the ability of the organism to cause infections; certainly it is associated with its ability to overcome the immunity that has been acquired during previous epidemics. The new epidemic strain is likely to have changes in its outer surface structure so that antibodies of people previously infected by cholera no longer recognise it. The centre of previous pandemics is often considered to be the Indian subcontinent so it might be expected that the causative bacteria of the next one will already be resistant to the antibiotics that are now commonly used to combat these epidemics. If the next pandemic is caused by resistant bacteria then it is going to be very much more difficult and probably very much more expensive to treat. When antibiotics are no longer active against cholera, unless a viable vaccine can be found, treatment will rely on infusion fluids. This will mean that most countries that have problems will have to obtain financial support from outside and this type of therapy can only be distributed through recognised medical centres.

It is becoming increasing clear that water-borne bacteria are a potential source of resistance genes. In a recent study of drinking water sources in India, there was high carriage of the genus *Aeromonas*. Members of this genus, particularly *Aeromonas hydrophila*, are associated with gastrointestinal infection, including diarrhoea, and they are increasing implicated in hospital infections. We have already seen with *Aeromonas salmonicida* that this genus can readily acquire resistance genes. My daughter, Alexandra, went to the sub-continent in 1996 and showed that the bacteria isolated in the drinking water sources in India were predominantly *Aeromonas sobria* (or *veroniae*); they were all resistant to amoxycillin and to the aminoglycoside streptomycin. Some were resistant to trimethoprim, nalidixic acid and cephalosporins; however, the most worrying is that some were resistant to the carbapenem imipenem which is considered to be the final defence against gram-negative hospital pathogens. The level of resistance to imipenem is sufficient to make these bacteria insusceptible to treatment either during gastrointestinal infection or if they cause hospital infection. *Aeromonas* species are not usually resistant to carbapenems and are often only so if isolated in hospitals, where they have experienced considerable pressure from the use of carbapenems. They often contain a β-lactamase that hydrolyses the carbapenems; however, this is usually very limited in its capability and, although present, usually does not confer resistance. Rachel Walker demonstrated the recently identified *Aeromonas* strains, from the water sources in India, the mechanism of resistance is quite new and unlike that found in previous *Aeromonas* strains. This represents a disturbing change in this pathogenic genus and may herald dangers ahead for the carbapenems in their ability to control *Aeromonas* species. The immediate problem in India is the emergence of an environmental bacterium that can cause disease that may soon be uncontrollable.

Both cholera and infections caused by the related *Aeromonas* species are preventable if the bacteria are not ingested. This means either severe chemical treatment or heat-killing of all drinking water. Chlorination is an effective method of water purification and can be used successfully in tropical areas as well as more temperate zones. All parts of Australia and the United States effectively have a pure water supply, even though they experience high temperatures and, in some areas, high humidity – conditions that will stimulate the growth of contaminating bacteria. The effectiveness of their water purification systems relies on maintenance of the equipment and the ability to clarify the water and remove all organic matter. In the developed world, water purification involves settlement tanks so that all particulates can settle and the clear water is then treated with chlorine. The chlorine will remove virtually all microorganisms except that it will not always remove *Cryptosporidia*. The addition of chlorine to water that still has particulates present, particularly the remnants of organic matter, is virtual annihilation of the capability to kill the remaining bacteria. This has been seen in some areas where chlorination plants are attached to town water supplies in India; the number of remaining microorganisms in the water is high because the chlorine cannot cope with the particulates entering the purification plant. Since the Indian government involvement in foreign commerce was lifted, American companies have returned to the subcontinent. In particular, the icons of American society, Coca-Cola and Pepsi-Cola, have instigated production plants in India. Both companies pride themselves that their products achieve the same degree of safety wherever in the world they are drunk. In India, this meant that both companies had to build their own water purification plants with the capability to settle the particulate material. The result is a product that does not always taste the same as it does in the United States but it is free of bacteria. It raised the question in India that if American companies can achieve this, why can the water supply companies not achieve the same? The answer is, of course, investment and the political will. The sewerage pipes and supply pipes are often close to each other and may be fractured, so may promote cross-infection. The water supply is often sporadic, particularly in the dry season, and without a continuous pressure within the pipes there is ample opportunity for contaminating bacteria to seep into the water supply pipes. This situation may be even worse in the monsoon, when the drainage is insufficient to remove the surface water, so it seeps through to the underground water supply, carrying with it any contaminating bacteria. The only method of ensuring pure water on a large scale and preventing contamination with sensitive or resistant bacteria, without total overhaul of the supply systems, is heat treatment, usually boiling, of all water

Alexandra collecting samples from the municipal drinking water supply in Vellore, India, in 1996.

for ingestion. In the country, there has been a policy of drilling deep bore-wells, which generally do provide water that is free of most bacteria but the water is often collected and stored in vessels that are themselves contaminated. The epidemiology of resistance, if not bacterial contamination, suggests that if we are to preserve our use of antibiotics in the developing world, cross-contamination in the community must be curbed and this is very difficult to achieve.

TYPHOID

Typhoid fever can also be a water-borne disease. It used to be rampant both in the United States and in Europe; more than 50,000 people in the United States died from typhoid fever caused by contaminated water in a five-year period from 1900 to 1904. When the cholera outbreak of 1973 emerged in Italy it became apparent that it was a relatively minor problem but it was also revealed that Italy had 10,000 cases of typhoid fever per annum. The causative bacterium, usually *Salmonella typhi*, is a member of the Enterobacteriaceae family of bacteria. *Escherichia coli* and *Shigella* species are in the same family and *Salmonella typhi* should respond to treatment with the same antibiotics used to treat these bacteria.

Typhoid is still a deadly disease despite the supposed capability to treat it and the development of vaccines. It still kills 700,000 people a year and this largely derives from poor hygiene, infection with large doses of bacteria and antibiotic resistance. It was considered, however, that the development of vaccines should eradicate *Salmonella typhi* as a serious pathogen. The early vaccines were unpleasant and could give severe side-effects, sometimes akin to the symptoms of typhoid itself. The development of the most recent vaccines has overcome that disadvantage and the vaccines can be taken without serious inconvenience. They are effective against low challenges of typhoid bacilli.

Salmonella infections usually require high infective doses in comparison with, for example, *Shigella* species. The infective dose may be a million cells or more. The majority of cells are irretrievably destroyed by the defence of the stomach acid. Anything that will reduce the impact of the stomach acid will lower the required dose for an infection to establish. If the patient is old or very young, the pH of stomach acid is raised. The closer the pH tends towards neutrality, the more bacterial cells survive. In addition, the ingestion of *Salmonella* bacilli with food will also reduce the destructive effect of the acid. It is considered that the majority of infections are initiated by contaminated water rather than associated with food. The dangers of uncooked food are generally recognised and it is rare for food to contain sufficient numbers of *Salmonella* bacilli for serious infection. The dangers of water are often less well considered and the volumes of water that have to be consumed in tropical climates favour the intake of large numbers of infecting organisms if the water is contaminated. Certainly there is nothing to dilute the stomach acid other than the volume of water itself. The vaccine will provide protection to increased levels of infecting bacteria so that a vaccinated person will, on average, be more capable of surviving a challenge of bacteria than an unvaccinated person. However, if the dose of bacteria is increased, then the vaccinated person is also likely to succumb. Vaccination serves to increase the threshold of bacteria required to initiate an infection, thus vaccination may prove little protection in areas where the challenge dose is particularly high.

The main control of the disease has fallen on antibiotics. The disease produces fever which may be reminiscent, in the early stages, of a flu-like condition. After an incubation period of 7 to 21 days, the illness begins with fever, lethargy, headache, and loss of appetite. Increasing weakness and abdominal discomfort develop during the second week, when a rose-coloured rash may appear. Intestinal bleeding or perforation may occur in the second or third week and can be fatal. The bacteria are transmitted from the faeces or urine of these patients and most often the main source of infection is from chronic carriers, who represent 2–5% of the patients who have recovered. In an untreated and unvaccinated population, the mortality rate of typhoid may be around 30%. The early administration of antibiotics can reduce the mortality rate to 2%.

Typhoid is a life-threatening disease so the therapeutic index does not have to be large. Thus antibiotics that might be too toxic for the treatment of common infections can be used with impunity against *Salmonella typhi*. When the side-effects of chloramphenicol were first recognised and less toxic alternatives were available, the World Health Organisation suggested that this antibiotic be reserved for the treatment of typhoid. The alternatives were bactericidal whereas chloramphenicol was merely bacteriostatic which, although it might be considered a disadvantage in some infections, appeared to make no difference in the treatment of typhoid. Thus, by the 1960s, typhoid treatment usually focused on chloramphenicol and the disease became easy to control, and the therapy was both effective and cheap.

Excessive use of the drug suddenly led to problems in Mexico. In the late 1960s large numbers of treatment failures were reported with chloramphenicol. Professor E.S. Anderson of the Enteric Reference Laboratory at the Central Public Health Laboratory identified that the bacillus had acquired resistance to chloramphenicol. This heralded a series of spontaneous emergences of chloramphenicol resistance in endemic typhoid areas. Why the bacillus had taken so long to become resistant remains unknown; it is possible that the chloramphenicol resistance gene, which usually encodes an enzyme that adds an acetyl moiety to the drug to render it inactive, does not survive well in *Salmonella* species. It is exactly the same gene as is found encoded by the plasmids of *Escherichia coli* and it thrives in this close cousin.

Alternative antibiotics were rapidly introduced to cope with the insusceptible bacteria. Unfortunately the alternatives could not be antibiotics reserved for typhoid, so trimethoprim (with sulphamethoxazole) and amoxycillin were extensively used. Resistance to trimethoprim soon appeared around the Persian Gulf and the resistance genes were related if not identical to those in *Escherichia coli*. In India, at the end of the 1980s, typhoid remained largely susceptible to chloramphenicol despite its widespread use throughout the country. Strains suddenly appeared spontaneously at various centres that were chloramphenicol-resistant. The resistant bacterium seemed to be exactly the same in each of the centres where it was examined closely, suggesting that it was the spread of an individual bacterium. At the start of the 1990s, the bacterium had started to acquire resistance to trimethoprim-sulphamethoxazole but still could be treated with amoxycillin, and amoxycillin soon became the drug of choice. As we have seen, the use of amoxycillin in the gut can provide the optimum conditions for the promotion of resistance and if the pathogen is able to breed in the gut, the environment for resistance acquisition is created. Within two years, almost all medical centres in India that had reported problems with chloramphenicol resistance were now facing a typhoid epidemic caused by a multi-resistant *Salmonella typhi* that had resistance genes to the three main drugs used to treat it. In fact it also had resistance genes to streptomycin and tetracycline but these would not normally have been used for treatment.

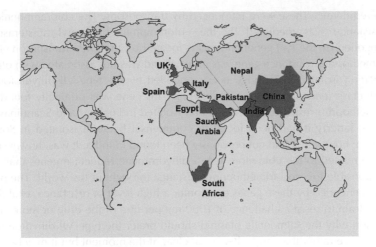

Countries where multi-resistant *Salmonella typhi* have been reported

A common plasmid was found in strains isolated from different parts of India; it was large (*c.a* 180 kb) and was part of the H incompatibility group. The epidemic looked as though the endemic *Salmonella typhi* strains had been infected with this plasmid and had become multi-resistant. Certainly examination of the plasmid DNA from strains isolated from diverse locations suggests that the plasmids were, if not identical, very closely related to one another. However, was this really an example of a plasmid epidemic or rather was a single strain spreading through the subcontinent? As molecular techniques improved, it became possible to identify individual bacteria by the fingerprint that their DNA made on digestion with restriction endonucleases. The technique, known as pulsed-field gel electrophoresis, cuts the chromosomal DNA with the restriction endonucleases, as is done with plasmids for their fingerprinting. However, the fragments are often of similar size and if separated by conventional electrophoresis would not separate adequately for a characteristic pattern to be determined. Therefore an electric field is applied at an angle (often 120°) and then applied at the same angle in the opposite direction. This allows larger DNA fragments to separate from one another. The conventional electrophoresis gives very characteristic fingerprints of plasmid DNA and pulsed-field gel electrophoresis gives a pattern that is equally diagnostic for chromosomal DNA. Pulsed-field gel electrophoresis gives very clear patterns with *Salmonella typhi* and demonstrated that, as far as can currently be determined, the outbreak in India has been caused by a single strain.

Why should a single strain suddenly become so successful? The plasmid must have something to do with it. It was able to capture resistance genes which were required when new antibiotics were introduced to combat the epidemic. These were introduced by transposons. The chloramphenicol resistance was mediated by the chloramphenicol acetyl-transferase I (encoded by the *catI* gene) and this is located on the Tn*21* transposon. The trimethoprim resistance gene was unusual and was not associated with other bacteria isolated from India. It was encoded by the type VII dihydrofolate reductase (encoded by the *dfrVII* gene). This gene was originally found in trimethoprim-resistant *Escherichia coli* isolated in England and Scandinavia. More latterly it had been found in commensal bacteria isolated in South Africa; however, it had not previously been found in India. It was, however, a gene related to the ubiquitous type I dihydrofolate reductase genes that are responsible for most trimethoprim resistance throughout the world. The proteins encoded by these genes can confer a high level of resistance, enabling the strain to resist a challenge of 1000 mg per litre of the drug or more. The reason why the salmonella plasmid should prefer the type VII dihydrofolate reductase rather than any other is not clear at the moment but it may be that its DNA has fewer sites susceptible to the salmonella restriction enzymes than the genes found more commonly associated with the plasmids of *Escherichia coli*. So the type VII gene may be the gene that had evolved specifically in Salmonella.

When the amoxycillin resistance gene was captured by the Salmonella plasmid, this was no species-specific resistance gene. The plasmid had been invaded by transposition of Tn3, the ubiquitous transposon that carries the TEM-1 β-lactamase gene (bla_{TEM-1}). The plasmid acquired the resistance gene that was already in the commensal faecal bacteria of virtually everyone in India; the sudden introduction of amoxycillin to treat typhoid simply created a disaster that was always waiting to happen because immediately, the commensal bacteria gave up their resistance gene to this, one of the most dangerous community pathogens. The abuse of amoxycillin for the treatment of trivial infections ruined any chance for it to control typhoid.

If the older antibiotics are now useless against typhoid, the newer drugs should do better, particularly if there is either very little resistance in community bacteria or the resistance genes are never located on plasmids. With this philosophy in mind, ciprofloxacin was introduced enthusiastically for the treatment of typhoid. There was no resistance to this drug and very little had been found in other community bacteria. The epidemic was controlled effectively and the number of multi-resistant bacteria started to reduce. Sensitive *Salmonella typhi* were found to be responsible for some of the newer outbreaks; these were not sub-populations of the multi-resistant bacteria but different strains of bacteria. The new strategy seemed to be working until there were reports of ciprofloxacin-resistant *Salmonella typhi* that had been isolated from infected travellers who fell ill on their arrival in England.

Soon isolates from India were showing decreased susceptibility to ciprofloxacin and there were reports of clinical failure. The bacteria had developed the traits that are associated with the emergence of ciprofloxacin resistance; the α subunit of DNA gyrase had undergone mutation, in particularly changes at positions 83 and 87, alterations that are characteristic of the development of ciprofloxacin resistance in other bacteria. The acquisition of ciprofloxacin resistance is never by plasmid and these chromosomal mutations are selected directly from the extensive and, perhaps, less effectual administration of the drug that has been discussed earlier in this chapter. The acquisition of ciprofloxacin resistance in *Salmonella typhi*, already resistant to the other antibiotics commonly used against it, is a disaster that may never be rectified. Ciprofloxacin represents the final defence against *Salmonella typhi*, and the inability to treat typhoid could, without rapid improvements in community public health and sanitation, cause a calamity responsible for millions of deaths.

Unfortunately there is no immediate solution. Immediate vaccination of the communities most at risk would stem the immediate risks but vaccination is not the final answer; treatment still requires antibiotics. There is no profit in the development of antibiotics specifically for the treatment of typhoid; the people at risk are those least likely to be able to afford the therapy. The price of new antibiotics usually has to placed at the top of the market in order to try to recoup some of the development costs; typhoid sufferers are not in a position to pay this.

OLD ANTIBIOTICS FOR POOR COUNTRIES

There is often a great deal of publicity surrounding the dumping of pharmaceuticals, which have failed safety tests in the developed world, in the developing world where cost is the major consideration. There have been concerns in the poorer countries of the Caribbean that sub-standard antibiotics are being sold there by the major pharmaceutical companies; this is unlikely to be the case and it is more likely that the drugs have been stored badly and thus reduced their potency. There is a lack of any firm evidence that sub-standard antibiotics are dumped on the developing world. The problem is rather the inability to obtain the latest and most powerful compounds.

In the summer of 1996, I travelled with my daughter, Alexandra, to the International Centre of Diarrhoeal Diseases in Bangladesh. The main problem there is shigelleosis which is now treated with antibiotics. While I was there, a working party of international delegates from the World Health Organisation were touring to make recommendations for the treatment of *Shigella* infections. The current recommendations are that if the bacterium looks to be resistant to amoxycllin and trimethoprim, which is a near certainty in many developing countries, then nalidixic acid should become the drug of choice. Nalidixic acid is an old antibiotic that never enjoyed much use in the developed world. It has the disadvantage that it cannot reach sufficient concentrations at most sites of infection. In the United Kingdom, it was used only to treat urinary tract infections because it was considered that, as the drug accumulated in the urinary tract as it is excreted through the kidneys, the concentration was high enough at this site to cure an infection. Nevertheless, the drug is hardly used as far better alternatives were commonly available. For this reason, the incidence of resistance remained low, but then the drug was used at approximately 1% of the rate of antibiotics such as amoxycillin or trimethoprim. It was promoted as having the advantage that there was no plasmid mediated resistance and thus if resistance increased it would not spread rapidly. This claim was never put to the test, as the limited usage of the drug did not trigger a significant resistance response. This does not mean that nalidixic acid use would not lead to resistance; in the study of commensal bacteria in India, resistance in *Escherichia coli* isolated from healthy people was nearly 30%, far higher than had ever been found before. Examination of the same bacteria for resistance to the more powerful quinolones, such as ciprofloxacin, revealed that there was none detectable at levels used to distinguish possible clinical failure.

The *Salmonella typhi* strains recently isolated in India are starting to show decreased susceptibility to ciprofloxacin. The same bacteria are highly resistant to nalidixic acid and are insusceptible to more than 128 mg per litre of this drug. This concentration of nalidixic acid will never be reached at any site of clinical infection. The extensive use of nalidixic acid in India has selected bacteria that are highly resistant to nalidixic acid and this resistance

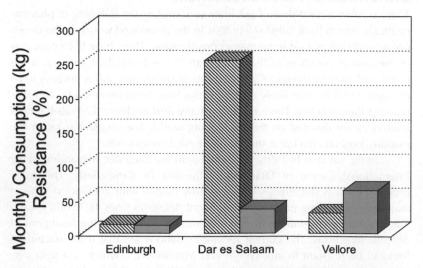

Monthly consumption of antibacterials trimethoprim resistance (%) in urinary bacteria

is the first stage in a series of mutations that will confer resistance to the more powerful quinolones. This will ultimately ruin the capability of ciprofloxacin to cope with these bacteria. The concentration of nalidixic acid in the gastrointestinal tract is insufficient to deal with infections so, for much of the time, the bacteria are challenged with sub-inhibitory concentrations. These are the ideal conditions for the emergence of resistance, particularly resistance mechanisms that derive from chromosomal mutation.

The consensus of many is that less powerful drugs should be used to treat diarrhoeal diseases before more effective versions are introduced. I submit that this could be a particularly dangerous strategy with nalidixic acid, despite the fact that it is inefficient at combating infections in the gut where it can barely reach sufficient concentrations; its capability to select resistance against itself and the fluorinated quinolones jeopardises the drugs that are likely to be used in the future. I asked the westerners in the working party if they, as travellers to the developing world, had brought any antibiotics with them to treat themselves if they succumbed to any of the diarrhoeal diseases. Not surprisingly, many travelled with full courses of antibiotics. None travelled with nalidixic acid but many had courses of the more powerful ciprofloxacin. The irony was that nalidixic might be suitable to recommend for use in developing countries but would never be tolerated in the developing world. If diarrhoeal diseases were common in northern Europe or the United States, there would be virtually no prescription of nalidixic acid; the more powerful fluoroquinolones would be used immediately. In southern Europe, diarrhoeal diseases are more common and are treated with antibiotics; they are virtually never treated with nalidixic acid but rather with ciprofloxacin. Nalidixic acid is cheap, particularly in comparison with ciprofloxacin, and the main reason for recommending it is cost rather than efficacy. The problem with allowing cost to drive the use of antibiotics is that it will almost certainly result in the introduction of weaker drugs which, in the case of nalidixic acid, is likely to lead to the erosion of the capability of ciprofloxacin.

WHAT IS THE THIRD WORLD?

The Third World is often defined as those countries that have unsustainable economies, often accompanied with pockets, often large, of the population who have insufficient income to sustain life. In terms of antibiotic usage, the definition of the Third World is very much larger. We have seen that there are parts of Asia and perhaps Africa and South America where antibiotic usage is unrestricted and resistance is out of control. The use of antibiotics has been inadequately controlled in many parts of Europe, particularly southern and now eastern, than it should have been. In many parts of Europe it has been possible to purchase antibiotics over the counter in pharmacies and antibiotics are administered freely when medical practitioners are consulted. The argument is that antibiotics are very widely used, so why should the general public not simply be able to purchase what they require over the counter without prescription? If the state has some form of socialised medicine, as most in Europe do, then it is much cheaper if the general population consult the pharmacist and purchase whatever is recommended.

This lack of regulation almost always results in increased resistance levels in community infections. There is no control on how the antibiotic is used and the tendency is that courses of antibiotics are split, particularly if they are expensive. The choice of antibiotic may be totally inappropriate and, at best, may result from a calculated guess of the problem by the pharmacist but at the very worst be based on the commercial pressure that a pharmacist has come under to sell a particular line of drugs.

There have periodically been questions raised in the United Kingdom as to whether the time has now come that at least some antibiotics should be freely available in pharmacies. Some general practitioners feel that they would be hassled less if those who "know" that they have a chest infection or a urinary infection prescribe their own therapy. Recently a friend of mine, without formal medical training, developed a severe cough. Presuming that he was suffering from bronchitis, he had access to antibiotics and administered fluoroquinolones to himself. His cough was a manifestation of lung cancer and the self-administration delayed the inevitable medical consultation. The lack of early medical attention, prompted by a belief in the ability to treat himself, probably led to a premature death. These fatal delays in consultation may or may not result from self-administration of antibiotics; a drastic increase in bacterial resistance is an inevitable result. Every country in the world that allows the self-administration of antibiotics has significantly higher incidences of resistance to all antibiotics that we have in the United Kingdom. At a time when few if any new antibiotics are launched for clinical use, the risks of increasing resistance are too high to allow the free availability of antibiotics.

It is difficult to determine whom the free availability of antibiotics would benefit. It is unlikely to serve the patients; they are certainly less experienced than the medical practitioner so will inevitably make a significant number of false diagnoses and certainly mis-prescribe the antibiotic. A patient is unlikely to be able to predict whether an infection is caused by a gram-positive or a gram-negative bacterium or even if it is caused by a bacterium at all. The patient, as the consumer, is also likely to be the ultimate victim as resistance increases and the antibiotic options decrease as resistance increases.

The medical profession might appear to benefit because there would initially be fewer consultations; if infections account for up 25% of medical consultations, this would be a massive potential saving, particularly in fund-holding practices. The benefit would be only short-term as patients who failed to treat themselves successfully would return, perhaps with bacteria that were far more difficult to treat than those that originally infected the patient. Those that had mis-diagnosed themselves so badly that they had allowed themselves to develop far more serious symptoms, such as a carcinoma or lymphoma, would cost the state far more as they undergo radical surgery. The devolution of antibiotic prescription away from the medical profession challenges one of the major roles of primary health care in the community, to monitor and control infectious disease. In Spain, Hungary, South Africa and the United States there is an epidemic of *Streptococcus pneumoniae*, the causative organism of a particularly virulent form of pneumonia. In Spain, the disease probably took hold because the monitoring of antibiotic resistance in the community is disregarded when antibiotics are freely available. This resistant strain has already been seen in some communities in England. There would little chance of monitoring its epidemiology or controlling its spread if many of the potential victims were prescribing penicillins as soon as they showed the symptoms of a chest infection.

The government, through the local health boards, might be seen to benefit if they could reduce the bill for pharmaceuticals and if they could reduce the sessions that the general practitioners had to work. The number of sessions might be out of the control of the local health board with fund-holders so this may not be a direct saving. Certainly the number of state prescriptions would initially fall. This is only a limited saving, as most antibiotics are priced little above the current prescription price and some actually cost *less* than the state prescription charge. The cost to the health boards of antibiotic treatment failures, as many are admitted to hospital with severe debility, particularly in chest infections, are likely to offset the savings made by the reduction in prescription costs. The savings could actually be negative and cost the health boards more.

Certain beneficiaries would be the pharmaceutical companies; the sales of antibiotics would increase markedly as the public gained access to what it considered to be more powerful remedies. The manufacturers of antibiotics usually only make substantial profits from drugs that are primarily intended for community use; the turnover for a single product can run into millions of pounds per day. The free availability of these drugs is likely to increase turnover considerably. If we presume that the sales would be restricted to pharmacies, then the pharmacists would benefit also. I have taught in a School of Pharmacy and the teaching of microbiology, which is adequate for the preparation of the antibiotic for the patient, is currently insufficient to deal with advice for each type of infection. The free availability of antibiotics is unlikely to benefit the patient and, if a state medical policy is designed to provide modern public health care, then increasing the access to antibiotics and removing the necessity for prior prescription by a medical practitioner has to be a retrograde step made for the basest reasons with few, if any, tangible financial benefits for the consumer.

WHAT ANTIBIOTICS TO TAKE ABROAD?

The British Army announced a few years ago that they had discovered a wonder drug that could keep infection at a minimum amongst troops who serve overseas. The wonder drug was ciprofloxacin, an antibiotic that was actually already widely used at the time by the medical profession and was commonly used to treat infections in the tropics! The announcement focused on the merits of travelling with antibiotics. Ground troops must travel with antibiotics, as their lives may depend on their ability to deal with a variety of infections quickly, but the general public should be more cautious.

Sometimes when talking to groups of medical practitioners I am asked what antibiotic I take abroad with me. In fact, I only ever take two. I take a course of metronidazole for anaerobic infections. This is not for general use but mainly to deal with deep tooth abscesses. A tooth abscess often means invasive dental procedures which could be dangerous if the equipment has not been sterilised adequately. HIV and hepatitis is common in Asia and Africa and I personally would not like to run the risk of infection because I had to seek dental treatment. So I always check that I have a dental check-up before travel and should infection occur, I can try to limit the infection. The other antibiotic is ciprofloxacin. I carry this for severe diarrhoeal infection or typhoid. Although I have been carrying it for nearly 10 years and I travel to the developing world each year, I have only had to use it once after a particularly unadvised meal in a sea-food restaurant in Penang, Malaysia. I have had my share of gastrointestinal hurry but it is usually so mild that liberal administration of pure water or sugary drinks has been all that is required for a rapid recovery. Antibiotics do not speed the recovery of most travellers' diarrhoea. These are usually caused by unfamiliar strains of *Escherichia coli*, to which the local population are immune. Antibiotics are an inefficient method of dealing with them, and avoidance is much more effective. The reason why infection is so high in the developing world is a lack of education. We do not have that excuse; we know what are the most likely sources of infections particularly those that are associated with travelling. I always ensure that the water that I drink is pure, usually boiled as I carry a portable kettle, and that all food is recently cooked and all fruit is peeled by myself.

Some people, including some eminent microbiologists, take antibiotics prophylactically. Before they travel, they start taking antibiotics and continue to do so the whole time that they are abroad. This is the same philosophy as anti-malarial prophylaxis where drugs that are used for malarial treatment are given before, during and particularly after treatment. The consequences of a western traveller succumbing to malaria are severe and thus prophylaxis is vital. It is recognised that it does provide a false sense of security and the other precautions to avoid malaria, including the use of nets and insect repellents, are often ignored. The use of prophylactic antibiotics has the same consequence; it provides a feeling of confidence that is usually unfounded and weakens the resolve to avoid possible sources of infection. During the conflicts in Malaya during the 1950s, it has been reported that the troops were given prophylactic penicillin to deal with syphilis and gonorrhoea that they might subsequently catch from the local prostitutes. Most travellers will take prophylactic antibiotics to avoid travellers' diarrhoea and this will often be seen to provide an opportunity to eat and drink with less caution. We have seen how typhoid is acquiring resistance to ciprofloxacin and the drug would be unlikely to cope with a massive dose, particularly if it had reduced susceptibility to the drug. The practice of prophylaxis is itself likely to increase the incidence of resistance because it means that more antibiotics are in use than are required to deal with infection. Most informed opinion suggests that antibiotic prophylaxis should never be used. If antibiotics are carried, they should be used cautiously and only after the more traditional remedies of oral rehydration have been tried and are not succeeding. If oral rehydration fails, there will be an antibiotic to fall back on. If the antibiotic fails, there may be no other remedy in reserve.

9

Antibiotic resistance in modern health-care management – why we shall lose the battle against infection

During the 1960s enormous concern was expressed at the possibility of antibiotic resistance causing treatment failures, particularly because of the use of antibiotics in animal husbandry. These problems were forecast to occur in community infections. In the late 1960s, a major weapon was developed in the fight against disease, transplantation. Christian Barnard's initial successes with heart transplantation were spectacular; one patient lived for months after the operation. All these transplant patients eventually died from rejection of the transplanted organ. The body's immune system recognised the new organ as the invasion of foreign tissue and tried to defend itself against this intrusion. The immune systems of transplantation patients had to be suppressed if they were to have any chance of survival. The patients were made immunodeficient, often with steroids. The same immune system is used to protect against bacterial infection so these patients also lost any defence against bacteria attack. To complement the immunosuppression, the patient had to be given a cocktail of antibiotics, given in far greater quantities than would be used against immunoproficient patients. This antibiotic treatment also departed from the philosophy of previous hospital administration of antibiotics; these drugs had to kill the bacteria (bactericidal). Previous hospital usage of antibiotics had relied on the simple prevention of bacterial growth and the leucocytes would remove the stalled bacteria (bacteriostatic). Once the bacteria had stopped growing there had never been any need to control them further, but these immunosuppressed patients required help to eradicate the bacterium completely. This had always been considered to be an impossible long-term goal for antibiotics, and without the help of the immune system the patient would be doomed.

In the 1970s, these pioneering surgeons were helped by a new generation of antibiotics which had recently been introduced. This advance was
initially based around the cephalosporins; the so called second-generation
cephalosporins were active against the hospital bacteria that had proved so
difficult to treat with the early drugs. The antibiotics killed most of the bacteria that lurked in Intensive Care Units and they could be used safely in large
quantities. This was in stark contrast to the aminoglycosides which had been
the mainstay of the control of severe hospital infection. These antibiotics
were so effective and safe that surgeons could give them prophylactically
during the operation to prevent the establishment of infection. There were
one or two bacterial species that were resistant to these antibiotics but they
were rare at the time; the major pathogens were gram-negative and usually
Escherichia coli or *Klebsiella* species.

By the end of the decade, resistance was causing difficulty in treatment but this was rarely from resistant variants of the bacteria that had previously infected these patients. New species of bacteria were now colonising
the immunosuppressed patients. Spectacular advances were also being made
in the treatment of cancers; however, the drug used to target the tumours
caused neutropenia, a deficiency of granulocytes circulating in the blood
which gives a concomitant decrease in the ability to resist infection. This can
occur in many diseases, most notably leukaemia, but is a particular problem
during X-ray irradiation or the use of toxic drugs during cancer chemotherapy. If there are fewer than 500,000 neutrophiles for every millilitre of blood,
the patient becomes very susceptible to infection.

In the 1980s, a third generation of cephalosporins based on the 7-
oxime ring substitution produced a group of compounds which seemed to be
insusceptible to all β-lactamases. Some of the most difficult pathogens,
including all species of *Klebsiella* and *Pseudomonas*, were treatable and hospital infection appeared almost completely controllable. However, although
these antibiotics were active against the gram-negative bacteria that were
prominent in Intensive Care Units and were most often responsible for post-
operative infection, they were virtually without effect against gram-positive
bacteria.

During this period of sensational medical advance, those who publicly
predicted that antibiotic resistance in clinical bacteria would rise until we
entered a "post-antibiotic era" where these drugs became almost totally ineffective were considered to be eccentric prophets of doom. The pharmaceutical companies appeared keen to meet the challenge and this was an exciting
time in the development of antibacterial drugs. The third-generation
cephalosporins were closely followed by the introduction of the fluorinated
4-quinolones, a completely novel class of antibacterials, and by the end of
the 1980s, the first carbapenem, imipenem, was developed. With this
armoury, surely we should have had sufficient antibiotics to last us well into

the next millennium and anyone propounding thoughts that resistance could ruin this marvellous dream must be a false prophet. The medical profession were, and still are, ignoring what is happening in their hospitals; with these drugs we have effectively shrouded our hospitals in an antibiotic blanket, resulting in continuous pressure to select resistance, and we are witnessing a progressive and sometimes alarming increase in resistance in nosocomial pathogens. This has resulted from both the acquisition of resistance pheno-types in pathogens that are normally sensitive and, perhaps of more concern, the emergence of inherently resistant bacteria which have not previously been considered serious pathogens.

The problem is now acute in Intensive Care Units but has been build-ing up progressively elsewhere in the hospital. The action of the latest gener-ation of antibiotics is poor against gram-positive bacteria, but the first of these problems was not concentrated in Intensive Care Units but rather amongst general hospital patients.

THE GOLDEN *STAPHYLOCOCCUS* AND ITS COUSINS

Ever since Fleming tried to cure infection, *Staphylococcus* was always con-sidered to be a problem. The *Staphylococcus* is a genus with a number of species; one of those species was considered to be pathogenic. *Staphylococcus aureus* was so called because it gives colonies of a charac-teristic golden colour. The difficulty in controlling this bacterium is that it is only pathogenic under certain circumstances. It is ubiquitous; many of us carry it on our skin or in our noses and it does us no damage at all. If, however, the skin is broken the bacteria can invade and produce a boil or carbuncle. If the wound is larger, perhaps after an operation, the *Staphylococcus* can cause a severe suppurative infection, which will proba-bly have to be drained.

Staphylococcus aureus has always been difficult to treat. When peni-cillin was first used against it, the bacterium rapidly used β-lactamases to combat the drug. To overcome the β-lactamase the semisynthetic penicillin, methicillin, was synthesised. Methicillin represented a major development in the battle against *Staphylococcus aureus*. The methyl group that was added at the 6-position of the penicillin nucleus increased the size of the molecule so that it was physically larger than penicillin G. The β-lactamases of *Staphylococcus aureus* are of the class A molecular group and the substi-tution provides the molecule with steric hindrance; it cannot enter the active site of the β-lactamase simply because it is too large. The molecule also has greater difficulty in inhibiting penicillin binding protein 2, and binds it at about 3% of the efficiency of penicillin G; however, the ability to resist the action of the β-lactamase more than offsets the reduced antibacterial effect. Methicillin became the drug of choice to treat *Staphylococcus aureus* throughout the 1960s. There were periodic reports of resistance but they did not seem important and, if the patient was severely ill, there was always the option to use gentamicin. Indeed, the initial reports of methicillin resistance were often thought to be overstated and the more widespread acceptance of gentamicin even resulted in a decrease of methicillin resistance.

In the 1980s a change occurred that, when we look back at the history of antibiotics, may be seen as the beginning of the end. The bacteria became gentamicin resistant and then, when these resistant strains were treated with methicillin, they became methicillin resistant. These bacteria have since been referred to as Methicillin-Resistant *Staphylococcus aureus* (MRSA) but it was not just the acquisition of methicillin resistance that signified the change, it was the concurrent acquisition of gentamicin resistance. The bacteria are much more accurately named by Ron Skurray as Multi-resistant *Staphylococcus aureus*. MRSA was the first of the resistant problematic bacteria in hospitals and the first to affect the successful management of infections in hospitals. The emergence of *Staphylococcus aureus*, resistant to both methicillin and gentamicin, occurred first in London and Melbourne. In some of the London teaching hospitals, this caused enormous problems in the treatment of some patients. In Melbourne, it resulted in the closure of wards. The press expressed trepidation about this "Golden Staphylococcus", claiming that it was impossible to enter hospital without succumbing to infection and, once infected, the patient should become a pariah. The infected patients did become hospital rejects; no hospital wanted them. Even now, the infected patient is an unwelcome admission to a hospital. This is surprising nowadays because almost every large hospital has its own problems with MRSA and the patients may not have been the cause. My mother was recently a patient in a major London teaching hospital, acquired MRSA and then became an infection control outcast.

The acquisition of methicillin and gentamicin resistance is just the start of a fishing expedition that the MRSA undertake to acquire as many resistance genes as possible. It does seem that the MRSA do have a predisposition to acquire new resistance genes and not just those associated with methicillin resistance; however, the mechanism of methicillin resistance is of particular interest because, although it may be aided by a β-lactamase, it is primarily manifested by alterations in the target site; the multi-resistant strains produced an additional penicillin binding protein, called PBP2', PBP2a. or PBP2'. This new PBP is likely to be a transpeptidase that can substitute for β-lactam-inhibited PBP in methicillin-resistant strains. As soon as they had acquired these two resistances, they became resistant to trimethoprim and other antibiotics as well as to some common antiseptics. In fact the bacteria acquired resistance genes to all suitable antibiotics except vancomycin. MRSA were originally considered to be spontaneous multi-resistant variants of the normal hospital *Staphylococcus aureus*; however, there have recently been major advances in molecular biological techniques that have improved

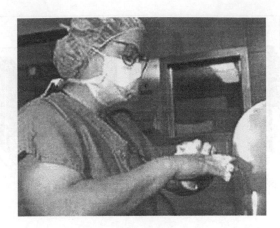

Hand washing – a major weapon in preventing the spread of resistance

the ability to distinguish bacteria. These techniques suggest that MRSA is not closely related to the resident hospital bacteria but rather that epidemic strains are spreading from one hospital to another. Close epidemiological analysis also suggests that rebuking the patient for introducing the strain is not fully justified. Recent work suggests that colonisation has to occur before an MRSA infection can occur. This colonisation can be on the skin of some but may well also be in the nose. It is suggested that this is how the bacterium could be introduced into the hospital; this could be carried in by a health worker as easily as by any patient. Regardless of the means by which it entered the hospital, it is likely that the principal mode of transmission through the hospital is by the transiently colonised hands of hospital personnel. One of the most difficult procedures to implement in hospitals is to persuade health-care workers to wash their hands between patients. Some Infection Control personnel, who despair at the inability to educate their colleagues into the necessity to wash their hands between the examinations of two patients, have now recruited the help of the patients. If the patient is well enough to speak, the patient is primed to observe the health-care worker's action from the previous examination. If they have not washed and decontaminated their hands, the patient is urged to state: "Why didn't you wash your hands after the last patient and before you examine me? I run the risk of being cross-infected by bacteria infecting the last patient you examined." Very few health workers can tolerate this type of humiliation if they have been challenged more than once, particularly if it is in front of their juniors. The point is that failure to wash hands between examinations and subsequent cross-infection of the second patient is straightforward negligence. In terms of the subject of this book, it has spread resistant bacteria far more widely than necessary but it is, of course, equally important to prevent the transmission of antibiotic-sensitive bacteria and viruses as well.

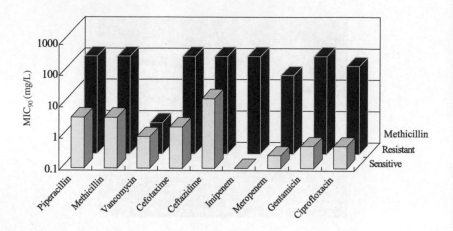

Data from Chanal *et al* (1989) J. Antimicrob. Chemother. **24 (Suppl A):**133.

Sensitivity of *Staphylococcus aureus*

We have had the problem of MRSA for nearly 20 years and, in Europe at least, its numbers still seem to be increasing. A recent study which covered the incidences of hospital bacteria in major hospitals throughout the continent showed that 12.8% of *Staphylococcus aureus* are multi-resistant, but this figure was not consistent; there was less than 1% in Scandinavia and more than 30% each in France, Italy and Spain. In this continent, the bacteria have now acquired resistance genes to all the major groups of antibiotics and are often only controllable with vancomycin. Although vancomycin resistance is transferable to *Staphylococcus aureus*, it has not been found clinically; which is just as well because if or when it happens, MRSA could well become uncontrollable.

Although *Staphylococcus aureus* is a traditional hospital pathogen, the coagulase-negative staphylococci were long considered far less pathogenic. These are the staphylococci that normally reside on the skin of us all. Their ubiquity makes them extremely difficult to control and like their more pathogenic cousins, some have become multi-resistant. The methicillin resistance is also manifested in exactly the same way as *Staphylcoccus aureus*, primarily by changes in PBP2. Also like *Staphylococcus aureus*, the acquisition of methicillin resistance is an indicator for multi-resistance and these bacteria can only be treated with vancomycin. These variants have become important opportunists because infections in immuno-compromised patients and in patients with indwelling prosthetic devices are often caused by hospital strains of multi-resistant coagulase negative staphylococci. They are much more difficult to eradicate than MRSA. In Western Australia during the 1980s there were major problems with MRSA. Although Western Australia is an

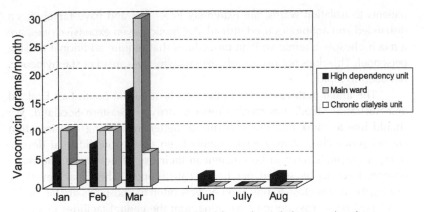

Brown, A.R, Amyes, S.G.B. *et al*. Epidemiology and control of vancomycin-resistant
enterococci (VRE) in a renal unit. *Journal of Hospital Infection* 40:115-124, 1998.

Effect of multi-drug resistance on the use of Vancomycin

area the size of western Europe, it has a limited population and it is largely
isolated from the rest of Australia and the world! This made it possible to
introduce a siege strategy whereby all incoming cases were carefully monitored and MRSA eradicated at source by expensive but effective barrier
nursing procedures. By the end of the decade, MRSA had effectively been
excluded from Western Australia but there were still massive problems with
multi-resistant coagulase negative staphylococci. They were never eradicated.

The routes of introduction into the hospital and then transmission of
coagulase negative staphylococci are probably similar to MRSA. When the
carriage rate was examined in Sweden, the proportion of the staff that were
colonised was quite low, around 20%. They mainly carried the multi-resistant coagulase negative staphylococci in their noses. Despite this, 82% of
the hospital clothing was contaminated so the probable mechanisms of
cross-infection were from contact by the hands and particularly the clothing
of the staff. Any member of staff with a nasal infection would also spread the
bacteria by air-borne and by direct transmission whereas the rest may simply
transfer the bacteria by indirect airborne transmission.

It is not clear why staphylococci should have mounted this final decisive challenge to antibiotics. Certainly it occurred at a time when hospital
budgets were undergoing serious reviews and the pressures were to reduce
staff and to increase the turnover of patients. Anyone who has been admitted
to hospital in the past few years will know that stays are much shorter than
they were even two decades ago. This will greatly increase the possibility of
cross-infection as the turnover of patients accelerates. The traditional
methods of controlling infection, either by barrier nursing or removal of the

patients to isolation wards, are extremely expensive and have largely been disbanded and antibiotics used instead. Antibiotics, even expensive ones, are a much cheaper alternative than procedures that require additional nursing personnel. This does not necessarily explain why the spread of staphylococci has been so rapid.

The initial outbreaks were in the new hospitals in South-East England and Australia; MRSA was much slower to arrive in eastern Scotland. We should like to think that this was due to better monitoring and infection control procedures. It may have equally been due to our hospital design. Modern hospitals have to be efficient in their use of personnel and more patients have to be served per health-care worker. This is achieved by placing the wards close to one another; very often the hospitals are tall buildings based on a cross design emanating from the central facilities including the lifts. The nursing stations may have ready access to all the wards on a single floor. Close monitoring of the staff may reveal that a significant number do not wash their hands before or after attending a patient. There is regular movement between the wards on a single floor and considerable movement between the wards on different floors. In this era of greater public access and virtually unrestricted visiting, the hospital regularly has hundreds of people who may be unwitting vectors of multi-resistant staphyclocci transmitting throughout the building.

The older hospital design places individual wards further apart; in Scotland it was called a pavilion design. The hospital was spread over a much larger horizontal area and the physical separation of the wards meant that the staff usually did wash their hands when walking from one ward to another. It did, of course, mean that more staff were required to run it. These hospitals were designed when infectious diseases were considered to be the major threat to life; however, this consideration is usually ignored in current hospital design as it is assumed that we shall always be able to control hospital infection. When we reach a stage where we can no longer control the spread of infection with antibiotics, those hospitals based on the older designs are likely to be better able to control the problem. We may see a return to the closure of wards or even hospitals that we first saw with MRSA infections 15 years ago.

Hand washing controls bacterial resistance and soap is an extremely good disinfectant, as it contains long-chain fatty acids which lyse bacteria. The surfactant action of soap ensures that bacteria lurking just beneath the surface can be eradicated. Soap is not, however, seen as a modern antiseptic and there are a variety of proprietary antiseptics available; many hospitals have their favourite brand, usually supplied with substantial discount from the supplier because of the quantity used. The staff are encouraged to wash their hands in the antiseptic before treating patients.

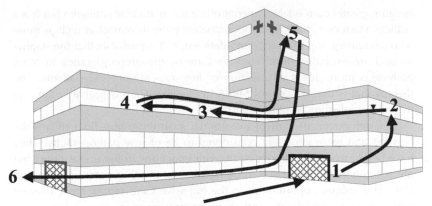

1. Entry into hospital via Accident and Emergency (ER)
2. Move to Pre-surgery ward
3. Move to Operating theatre
4. Move to Intensive Care unit for recovery
5. Mover to General Post recovery ward
6. Exit from hospital perhaps to convalescent home

Minimum number of moves of an emergency surgical patient admitted to hospital

The previous chapters have outlined the careful monitoring of antibiotic resistance throughout the world, but there is virtually no monitoring of susceptibility to the antiseptics used in the hospital. Most proprietary antiseptics are far more effective against gram-negative bacteria than they are against gram-positive. This will ensure that the antiseptic policy is providing an environment for gram-positive bacteria to proliferate. Much more concerning is that the multi-resistant *Staphylococcus aureus* and coagulase-negative staphylococci, during their fishing expedition to acquire as many resistance genes as possible, have imported genes that inactivate many antiseptics. The pathogens that many antiseptics are used to control are already *resistant* and as most hospitals do not test sensitivity to antiseptics, they are ignorant of this. The antiseptic policies may actually be selecting multi-resistant bacteria. This is borne out by the fact that the substantial increase in the antiseptics and disinfectants used appears to have very little controlling influence The resistance genes of MRSA and the multi-resistant coagulase-negative staphylococci are responsible for their current invasion of our hospitals and we should now consider that they are, therefore, a major contribution to their pathogenicity.

ENTEROCOCCUS
Staphylococcus aureus is a pathogen of the general hospital wards; despite its prevalence it is found less often in Intensive Care Units. This may be

because greater care is taken in controlling this particular pathogen but it is a pathogen that can attack immunocompetent patients almost as well as those who are undergoing some immuno-deficiency. It is probable that the staphylococci are excluded from Intensive Care by the preponderance of other pathogens more adept at infection of immuno- suppressed patients. The therapy that is used to control MRSA in Intensive Care is probably the gate that allows infection by other pathogens.

Enterococci are part of the normal gastrointestinal flora of healthy humans and were generally considered to be of low pathogenicity. They have been associated, on occasion, with infections in the urinary tract but are not generally considered to be pathogenic in other potential infection sites. The unique opportunities in the Intensive Care setting have allowed enterococci to proliferate and they are now seen as the second most common cause of nosocomial infections in the Intensive Care Unit. It was always assumed that the patients infected themselves with enterococci from their own gastrointestinal tracts; however, like multi-resistant staphylococcal infections, epidemic strains of enterococci are now found to be the major infectious agent. This conclusion has been achieved by recent advances in the molecular typing of these strains and they have highlighted the importance of cross-infection as the major disseminator within hospitals.

These recent observations have serious implications because, if the patient's commensal bacteria were reinfecting the patient, this would be an unfortunate but tolerated circumstance. It would be virtually impossible to demonstrate and unreasonable to suggest that poor medical practice would be responsible for this type of self-infection. If, on the other hand, the patient becomes infected with a hospital epidemic strain, this may have legal implications, particularly if it can be proved that the patient acquired this infection through some form of negligence by one of the medical staff.

Hospital patients have been carriers of enterococci for centuries but it is only in the past few years that they have been recognised as a problem. Unlike staphylococci, they have not been seen as lurking pathogens. Enterococci do have an inherent resistance to many antibacterials and it is this that has promoted their emergence as major nosocomial pathogens. This has promoted their selection and survival under the antibiotic blanket.

The recent introduction of the fluoroquinolones and later generation β-lactams has demonstrated that enterococci have the capability to respond by the rapid acquisition of new resistance traits. The genus has always had low-level resistance to penicillins, cephalosporins and aminoglycosides but this was soon extended with the appropriation of resistance to tetracyclines, trimethoprim and chloramphenicol. The capability of resistance to penicillins, aminoglycosides and clindamycin was increased so that high levels of these drugs could be resisted.

As more enterococci have been found in patients in Intensive Care Units, the control of multi-resistant enterococci has depended on treatment with glycopeptides, principally vancomycin and the fluoroquinolones. Ciprofloxacin is not usually considered as a drug of choice for many gram-positive bacteria but, *force majeure*, it had to brought into the armamentarium against enterococci. This inevitably led to an increase in resistance to fluoroquinolones amongst enterococci.

The final defence against multiply resistant enterococci has been vancomycin. It is the emergence of resistance to this antibiotic that heralds the brutal truth of untreatable enterococcal infections. Resistance to vancomycin has now been demonstrated to be very complicated. There are certainly epidemic strains that spread rapidly through hospitals but the resistance is manifested mainly by two resistance genes, *vanA* and *vanB*. These are both transferable genes and they are located on plasmids or conjugative transposon. The *vanA* resistance mechanism is the most worrying, not just because it is plasmid-mediated but because it also confers resistance to the other glycopeptide agent teicoplanin. It also confers a much higher level of resistance to vancomycin, equipping the strain with the capability to resist 1000 mg per litre or more. The *vanB* resistance mechanism is also transferable but usually confers a lower level of resistance to vancomycin and none to teicoplanin.

In the classical model of antibiotic resistance spread, it might be suggested that these two resistance genes rapidly disseminate through the clinical enterococcal populations. Certainly there is some spread of the resistance genes into different enterococcal strains and this is presumed to be by direct plasmid transfer. However, in the Royal Infirmary Edinburgh, there has been an outbreak of vancomycin-resistant enterococci and, although there has been some transfer of resistance plasmids, it has been dominated by the spread of a single strain of vancomycin-resistant *Enterococcus faecium*. The clonal spread of multi-resistant enterococci as the likely cause of an outbreak is supported by the predominance of multi-resistant *Enterococcus faecium* amongst the multi-resistant enterococci. This is in stark contrast to the sensitive enterococcal populations where *Enterococcus faecalis* are far more prominent. This suggests that multi-resistant *Enterococcus faecium* are able to spread much more easily in an antibiotic-rich environment.

We are now faced with an impossible problem; when the multi-resistant outbreak occurred, we simply did not have the antibiotics to eradicate it. Vancomycin was predominantly used to treat enterococci, staphylococci and clostridia and we had to restrict its use. The infection control team introduced barrier control measures, minimal vancomycin usage was permitted and alternative antibiotics were sought for control of clostridia. This type of strategy demands that MRSA be controlled without antibiotics for fear that additional vancomycin would exacerbate the problem, and unwittingly gives us a vision of the future when this strain becomes vancomycin-resistant. We controlled the outbreak, which were really a series of outbreaks as it was made up of a combination of clonal spread of *Enterococcus faecium* carrying *vanA* and a myriad of *Enterococcus faecalis* and *Enterococcus faecium* into which the mobile genes of *vanA* or *vanB* had invaded.

The greatest threat is if this plasmid-mediated glycopeptide resistance spreads to strains of methicillin-resistant *Staphylococcus aureus* as this may certainly herald a post-antibiotic era and mark the return of the hospital staphylococcus of old. The plasmid transfer of glycopeptide resistance from *E. faecalis* to *S. aureus* has been demonstrated under laboratory conditions so it would appear that there is no barrier to this occurring in nature. Complacency had cited that this transfer has never occurred in nature but vancomycin resistance has already emerged in methicillin-resistant *Staphyococcus aureus*.

There are recent reports, by Professor Hiramatsu, of MRSA in Japan that have MICs of vancomycin of around 8 mg per litre which are certainly sufficiently high enough to predict treatment failures. The strain is characterised by a cell wall that is twice normal thickness and increased production of penicillin binding proteins and murine precursors compared to those of vancomycin-susceptible MRSA strains (MIC≤2 mg per litre). The doomsday scenario of the *vanA* or *vanB* genes transferring to *Staphylococcus aureus* from the enterococcus has not yet materialised but this emergence in Japan is worrying enough and might be the start of a major epidemic. Microbiologists enjoying trying to second-guess what devices bacteria might use to overcome antibiotics and often they get it wrong. We presume that the successful resistance genes of enterococci must be the mechanism that staphylococci use to resist vancomycin. It is more than likely that staphylococci will employ a method other than the one used in enterococci and thus every new mechanism must be taken seriously and investigated.

Most hospitals are unable to eradicate vancomycin-resistant enterococci and they continue to spread. The number of hospitals harbouring these bacteria continues to increase and there are now few left that are completely free of them. Twenty years ago they were not considered of any significance within hospitals, but their prevalence as nosocomial pathogens and our inability to use antibiotics to treat them ensure that vancomycin-resistant enterococci must be considered to be an immediate crisis and we do not look as though we have any solution in the foreseeable future.

PNEUMOCOCCUS

Streptococcus pneumoniae is not considered to be a major hospital pathogen but it is relevant to consider with the previous two gram-positive genera. It is the major causative organism of pneumonia and certainly gives the severest form except for those HIV-positive patients infected by *Pneumocystis carinii*.

The pneumococcus can also be responsible for meningitis and, more mildly, for Otitis media. The most popular and successful treatment has been penicillin and for decades the pneumococcus has remained exquisitely sensitive to penicillin. In 1977, bacteria emerged which were less sensitive to the β-lactam antibiotics. It is not certain why this switch occurred so suddenly but it may be related to the introduction of cephalosporins with extended activity. The pneumococci have no ability to produce β-lactamases and, because they are gram-positive bacteria, have almost no opportunity to set up any permeability barrier to antibiotics. However, it was shown by Fred Griffiths in the first half of the twentieth century that pneumococci were able to transform whole sections of DNA. This meant that if there was free DNA in the vicinity of the pneumococci bacteria, they were able to absorb it. If the DNA was similar (greater than 65%) to the DNA of the pneumococcal chromosomal then a recombination could occur. The foreign DNA could swap with the equivalent part of the indigenous chromosomal DNA and become part of the genetic material of the bacterium. Although many bacteria can perform this type of transformation, it is particularly efficient in the pneumococci. It is thought that it evolved to allow the pneumococci to absorb genetic material that would provide the opportunity to change its antigenic structure to deceive the immune system. The pneumococcus is able to use this device to produce resistance to antibiotics.

The targets of the β-lactam antibiotics are the penicillin binding proteins, and alterations in these are the only opportunity that pneumococci have to overcome β-lactam attack. They can incorporate DNA, from other

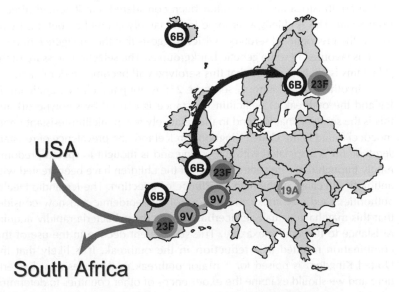

Spread of penicillin-resistant *Streptococcus pneumoniae* through Europe

bacteria, that encode penicillin binding proteins. When this DNA enters the cell, a recombination may take place and insert a large section of DNA into one of the penicillin binding protein genes. This new section of DNA forms a mosaic gene and may encode an alteration that would resist the binding of the β-lactam. In *Streptococcus pneumoniae* there are six PBPs and decreased affinity of PBP1A, PBP2X and PBP2B together give high-level penicillin resistance. Changes that might be selected by cephalosporin usage to provide resistance focus on decreased affinity of PBP1A and oxacillin would select changes in PBP2B. However, resistance to almost all β-lactams also requires decreased affinity of PBP2X. In clinical isolates alteration of PBP2X is achieved by the acquisition of whole blocks of nucleotides from related species of streptococci that are part of the normal oral flora; the most likely donor is the PBP2X gene from *Streptococcus mitis*, a bacterium sometimes associated with tooth decay.

The trick that *Streptococcus pneumoniae* uses to make itself unrecognisable to the immune system is to change its antigenic structure; this is achieved by some considerable genetic changes largely achieved by importing foreign DNA. As a result, there are many different serotypes of *Streptococcus penumoniae*. Some of these serotypes are associated with infections in children and others are found to be responsible for infections in adults. There are predominant serotypes within both groups but the most predominant species usually remain sensitive to the β-lactams. In Spain, one of these serotypes, 23F, emerged as penicillin resistant and then disseminated. This serotype has also been found to be predominant in the United States and in South Africa and it has often been considered that it spread directly from Spain. This is chiefly a serotype found largely in children but it is by no means the predominant serotype, which suggests that the emergence of resistance is favoured in some genetic backgrounds. The selection pressure of the β-lactams is likely to ensure that this serotype will become predominant.

In other parts of Europe, serotype 23F is not particularly significant. In Iceland the outbreak of penicillin resistance is caused by serotype 6B and this is the serotype that is found in the relatively few penicillin-resistant pneumococci found in the United Kingdom. In Iceland, the prevalence of resistant serotype 6B is associated with the young and is thought to spread predominantly through day-care centres. Many of the children have been treated with antibiotics because of the high incidence of infection. The Icelandic Health authorities used co-trimoxazole to control this epidemic and now consider that this may have actually exacerbated it, as these bacteria rapidly acquire resistance to this co-trimoxazole. The subsequent decline in the use of this combination resulted in a reduction in the outbreak. It is likely that the United Kingdom is poised for a major outbreak of these resistant bacteria here and we should examine the experiences of other countries to determine what therapy would best limit the damage.

When the bacteria were first found the level of resistance was not high and they have subsequently been classified as intermediate resistant. Strains that were able to resist higher levels of β-lactams subsequently emerged and these were classified as fully resistant. In the Spanish outbreak, this has been well defined: the penicillin-sensitive strains which are susceptible to 0.06 mg per litre penicillin, the penicillin-intermediate resistant strains which are resistant to 0.006 mg per litre but susceptible to 1 mg per litre, and the penicillin-resistant strains which are resistant to 1 mg per litre. Unfortunately, the increase in penicillin insusceptibility is accompanied by increases in resistance to almost all other β-lactam antibiotics. The penicillin-intermediate-resistant strains show a 10-fold decrease in susceptibility for all β-lactams; the highest degree of resistance is against the third-generation cephalosporins, particularly ceftazidime. Decreased susceptibility is seen against the carbapenems, imipenem and meropenem.

Treatment options still remain for resistant pneumococci and bizarrely may focus around β-lactams, particularly penicillin. The level of penicillin that is reached in the infected lungs or the middle ear is likely to be sufficient to kill bacteria of intermediate resistance. If the local surveillance is good then, in the absence of highly resistant strains, the continued use of penicillin may succeed provided sufficient concentrations are used. More difficult decisions have to made for meningitis because the causative organism will not have been identified before therapy has had to be effected. Treatment with an antibiotic to which the causative bacterium is resistant could result in death.

Why should this problem be so great in the pneumococci? Respiratory tract infections account for approximately 22% of all antibiotic prescriptions and there has been a heavy reliance on the β-lactams and recently on the cephalosporins. If the progression through sensitive to intermediate and full resistance is examined, there is approximately a 10-fold increase in the resistance to penicillin at each stage. Sensitive bacteria might have a median MIC of 0.02 mg per litre. Bacteria of intermediate resistance might have an MIC of around 0.2 mg per litre whereas resistant bacteria might just be able to overcome a concentration of 2 mg per litre. If resistance to a spectrum of β-lactams is examined, there is approximately a 10-fold increase in resistance at each stage. This increase is a little lower for some of the carbapenems which suggests that these β-lactams remain robust in the ability to overcome pneumococci but, more importantly, it suggests that they were not responsible for the selection of resistance. The proportional increase is greater for some of the cephalosporins, particularly ceftazidime and cefixime, which suggests that resistant bacteria may be more capable of overcoming these β-lactams. It would also support the view that the emergence of these resistant bacteria coincided with the introduction of extended-spectrum cephalosporins. Some have suggested that calling these bacteria penicillin-resistant pneumocci is a misnomer and that β-lactam-resistant might be more suitable; however, like the mis-named MRSA or VRE, these are multi-drug resistant bacteria.

When extended-spectrum cephalosporins were developed, it was impossible to provide them with the capability to pass through gastric acid without decomposition, thus they could not be given by mouth and were usually administered parentally. This problem has now been solved and some are now available for oral use. This makes them much more accessible for use in the community and it could be expected that resistant penumococci will become a greater problem. Multi drug-resistant *Streptococcus penumoniae* is destined to become the greatest infection problem in the community and we must ensure that we have the optimum surveillance measures in place to detect an increase in resistance and that we can attribute that increase to an identifiable strain of known serotype. It also raises questions about the wisdom of releasing oral extended-spectrum cephalosporins into the community; there are very few indications where these drugs might be required and there are usually plenty of suitable alternatives.

Methicillin-resistant *Staphylococcus aureus*, vancomycin-resistant *Enterococcus faecium* and *Enterococcus faecium* and penicillin-resistant *Streptococcus pneumoniae* form an elite group of gram-positive bacteria that are really all part of the same trend. The respective resistances that are attributed to them mark the turning point when the bacteria developed resistance to a target antibiotic that was always considered to be a safe bet. In fact, they are all multi-resistant bacteria and there are few if any suitable antibiotics to treat them. The methods by which they have developed multi-resistance may differ, the pneumococci transforming new genetic material while the staphylococci went through a series of mutations. They have been selected by our current armamentarium of antibiotics and we are largely seeing the clonal spread of epidemic strains. Our traditional view of the rapid spread of bacterial resistance is by horizontal gene transfer but, in these particular cases, we are seeing the selection of multi-drug-resistant strains and then rapid dissemination. We have seen that they are all resistant to the extended-spectrum cephalosporins which were introduced into clinical practice during the 1970s to combat what was seen as the greatest curse in hospitals, the gram-negative bacteria that caused infection in Intensive Care Units. The fluoroquinolones were introduced in the 1980s and then used extensively both in hospitals and for the treatment of chest infections. They have very poor activity against gram-positive bacteria and they are either already inherently resistant or resistance develops to them very quickly. Their extensive use is likely to have promoted the selection of these elite multi-resistant bacteria.

MULTI-DRUG-RESISTANT TUBERCULOSIS

We had really considered that tuberculosis had been eradicated. Whereas this might have been true in the developed world, at least 1% of the population of the Indian subcontinent, at any one time, were suffering from the clinical manifestations of this bacterial infection. During the 1980s, there was a sudden rise in tuberculosis in the United States but unlike the bacteria isolated before the Second World War, this organism was multi-resistant.

The aetiology of tuberculosis is not straightforward. It always used to be a disease that attacked the young. While the child was breast-fed, it was receiving protection against infection through the colostrum. When the child is weaned, this protection is removed and the child has no major protective antibodies. Thus throughout childhood, the individual is most at risk, for when it reaches adulthood, it may have acquired some immunity by subclinical exposure. Pulmonary tuberculosis is thought to produce symptoms in only 6% of those infected. This would mean that the vast majority of the population are protected in some way. As the individual reaches old age, the immune system becomes less efficient and there is greater susceptibility to infection, including tuberculosis.

The only population protection that we have is immunisation with BCG. This has been taken from a Mycobacterium strain that was isolated from a cow in 1902. It was then serially grown in a beef extract medium for 13 years. It was transferred 231 times at three-weekly intervals and by the end the bacterium was no longer pathogenic. Subsequent analysis shows that it has lost some of the genes necessary for pathogenicity and this can be inoculated directly as a live vaccine; indeed it does not work effectively unless it is alive. BCG immunisation is thought to give the individual protection for approximately five years but its main role is to provide protection for the human herd. If children in their early teens are given this vaccine, they

George Orwell died from an untreatable streptomycin-resistant strain of *Mycobacterium tuberculosis* on 21 January 1950

Antibiotic-resistant tuberculosis victim

will themselves be protected but they will also not be introducing this bacterium back into their family homes and thus infecting their younger siblings. These are the group that are at greatest risk and they have to be afforded some protection; however, we have until recently kept our very young in individual homes until they are ready for school. The modern trend is, because both parents work, to place children in day-care centres from a very early age and the opportunity for infection to occur is considerably increased.

The BCG vaccine is predominantly to guard against paediatric infection and does not generally protect in adult life, where is assumed that some self-protection has been attained. In the developed world, the living conditions of the middle classes do not favour the spread of the organism. In Britain during the 1930s, the number of rooms within a household overtook the number of occupants. This has meant that for 60 years, the majority of children have had their own bedrooms and certainly their own bed. This is a significant barrier for the spread of tuberculosis. In the developing world, the exposure and living conditions still favour the spread of the bacterium and BCG might give only limited protection.

The reason why tuberculosis arose again in the United States during the 1980s was not due to the failure of the vaccination programme but to two basic breakdowns in health-care management. In some cities, they had stopped monitoring the number of cases of tuberculosis; it was considered to be a disease in rapid decline and was forecast to be extinct by 2010. This measure failed to recognise the growing numbers of urban poor in the large cities. These groups could not afford health care and many must have failed to receive medication but they would not be included in many statistics. There was a sharp decline in the number of cases in the middle classes and this group might have been predicted to lose tuberculosis as an invading pathogen. The other cause of the resurgence of tuberculosis would affect the middle classes. Tuberculosis is a disease of attrition between Man and microbe; the bacterium is always present, but Man survives because of immunity. In the 1980s, the emergence of the transmissible Human Immunodeficiency Virus would upset this balance. As HIV infections progress to the full symptoms of AIDS, the patient loses protection to mycobacterial infections. Even if the patient is an adult, he or she may lose protection to the strains that predominantly infect the young. The only opportunity to protect the patient is to give aggressive antibiotic therapy. Normally when antibiotics are given to a tuberculosis sufferer, the bacteria that are actively growing are inhibited and killed within the first week of treatment. The treatment usually continues for a further six months because the bacterium is intracellular and some cells may be difficult to reach; also, it may be in a dormant stage and continuous therapy will kill any bacteria as they are reactivated. In the end, the otherwise healthy individual will eradicate the remaining bacteria by their immune system. The AIDS sufferer has a compromised immune system that cannot take over this function and the patient will have to be on much more aggressive antibiotic regimens. This will inevitably mean more antibiotics for a greater period of time, in other words a much greater selective environment for the emergence of resistance. This can be seen in the siting of the wards in some modern hospitals. There have been incidences where the tuberculosis wards and HIV wards are side by side; after all, they are both infectious diseases and would be treated by infectious disease physicians. In this scenario, the previous uninfected HIV patients can rapidly acquire pulmonary tuberculosis. In the old British fever hospitals, the wards for different infections were often in different buildings and the opportunity for this type of cross-infection would be very much reduced.

We have already seen that the emergence of resistance in tuberculosis is never by horizontal gene transfer but by the accumulation of mutations in the genes encoding the respective drug targets. It does not matter whether the bacteria are isolated from immuno-compromised or competent patients or where in the world the bacteria come from, the mutations are usually the same. The first-line drugs against pulmonary tuberculosis are isoniazid, rifampicin, streptomycin, ethambutol and pyrazinamide. These drugs have the advantage that they are able to kill the bacterium but they have been available to us for nearly 50 years. The alternative second-line drugs are ethionamide, kanamycin and the fluorinated quinolones, which are also bactericidal but less effective. The third line are para-amino-salycilic acid, cycloserine and capreomycin. These are weak drugs and would be ineffective over a long period of time, especially for AIDS patients as they are only bacteriostatic. There is now resistance to each of the first-line drugs. Isoniazid is a prodrug and has to be activated by catalase (peroxidase) within the bacterium so that it can inhibit the mycolic acids that make up the cell wall. Resistance can simply be achieved by a mutation that reduces the catalase activity and the prodrug cannot be converted to the active inhibitor. There may also be hyperproduction of the *inhA* gene product; this will mean a surfeit of active sites to bind the converted drug and this would inevitably reduce the capability of the isoniazid to inhibit cell wall production. A similar mutation rate will provide resistance to pyrazinamide but the exact mechanism is unclear. Pyrazinamide has an unknown action and thus alteration to the target site cannot be measured. However, it is believed that pyrazinamide is not the active component and, like isoniazid, has to be modified once it enters the cell. The mechanism of resistance has been speculated to be similar to that of isoniazid, an inability to activate the drug. Resistance to streptomycin, rifampicin and ethambutol is essentially by the same mechanism; the binding site of the active drug is altered so that it is no longer inhibited by the drug. There will inevitably have to be some compromise and the altered target is usually not as efficient as the original. This is often seen with alterations of the *rpsL* gene product where the binding protein on the smaller ribosomal subunit is altered so that it no longer binds the drug; in fact this is very efficient and can increase the resistance of the host by well over 1000-fold. It is often argued that these changes should disable that bacteria so that they have reduced capability to produce infection. This does not appear to be the case with tuberculosis for the multi-drug-resistant strains appear to spread as readily as their sensitive counterparts.

Multi-resistant tuberculosis does appear to be declining but there is still sufficient in some communities to remain a cause of concern. Why should there be a decline? We certainly manage our HIV-positive patients more successfully than before. Infection control measures should now prevent them from contracting the disease when coming into hospital for treatment. In the developed world, more public awareness of the disease leads to greater compliance with therapy even though it is spread over many months. The modern regime is to administer four antibiotics on diagnosis; the bacteria are then cultured in the laboratory and sensitivity tests performed. Approximately one month later, the patient has the therapy reduced to two drugs. If the bacteria are resistant to more than two drugs in the original therapy, new drugs are introduced into therapy at this stage though they are often less effective.

Mankind is extraordinarily lucky that tuberculosis should be a relatively primitive organism in terms of its ability to disseminate resistance. Almost all other major pathogens are able to transmit resistance genes from one bacterium to another but this ability is missing in the mycobacteria. The insertion sequences associated with transposons are present in mycobacteria but plasmids seem incapable harbingers of resistance and no plasmid transfer of resistance genes has ever been found. We simply do not know the reason why but we should be grateful. Plasmid transfer amongst the mycobacteria would soon render the multiple-therapy regimens useless and we would be faced with a pandemic of untreatable tuberculosis. Nevertheless, we desperately need new drugs; no new drugs have been developed for over 10 years. We need drugs that should be long acting, perhaps by a slow-release mechanism or at least with long half-lives. Tuberculosis is primarily a disease of poverty and deprivation so those infected are often not capable of paying for the therapy that they need. This is a disincentive for a commercial pharmaceutical company investing much time and effort in research into new anti-tuberculosis therapy. Perhaps the greatest hope is to find new vaccines or even immunotherapy. Whatever the solution, tuberculosis is a global problem and its control will always have to be paid for by the richer societies, who are usually less affected by it. It is one of the major public health problems of the world and requires a global strategy to control it.

CEPHALOSPORIN-RESISTANT *KLEBSIELLA* AND *ENTEROBACTER*

If the introduction of penicillin G made the greatest impact on our ability to control infectious disease, the closely related cephalosporins have had the greatest effect in the control of hospital infection. Until their introduction, serious infections had to rely on the toxic aminoglycosides and hospital infection had to be monitored closely. The development of cephalosporins, especially the so-called second generation, ensured that hospital infection could be limited with a relatively safe drug. It meant that little consideration had to be made as to whether the advantages of therapy outweighed the risks; there were virtually no risks. Cephalosporins could be used prophylactically in surgery and patients were often given courses of antibiotics to provide cover against infection during operations. This gives the surgeon a greater margin of safety and certainly reduces the incidence of post-operation infection. It did result in a massive increase in the exposure to antibiotics and thus the potential for the selection of resistance increased. Often during the recovery period, particularly if the patient developed an infection during intensive care, cephalosporins would be administered. The most common pathogen in intensive care was *Escherichia coli* which usually responded well to cephalosporins. They were even successful at controlling the *Klebsiellae* which did not respond to the pencillins. Thus the use of antibiotics increased and an almost total reliance was placed on cephalosporins as the general workhorse to control gram-negative infections in hospitals. Some problems were experienced with infections caused by *Enterobacter cloacae*, a cousin of the *Klebsiella* genus. Enterobacter strains have an indigenous β-lactamase whose production is normally rigidly repressed because its production would weaken the strain and reduce its competitiveness. The bacterium would normally switch β-lactamase production on if challenged with a cephalosporin. However, in an environment of heavy cephalosporin use, this is a slow response and often insufficient to overcome the attack. The continuous production of the β-lactamase provides the bacterium with a massive advantage and the selective pressure on these bacteria favours those that have a mutation in the mechanism that represses the β-lactamase production. The stably de-repressed (constitutive) mutants only have an advantage when the antibiotic selection pressure is high; otherwise the hyperproduction of the chromosomal β-lactamase is a massive drain on the resources of the cell. The de-repressed mutant is at a considerable disadvantage when the antibiotic pressure is removed. However, despite this disadvantage, these mutants are very prevalent, particularly in Greece where they represent more than 60% of all *Enterobacter cloacae* strains isolated. We are not even exempt in the United Kingdom, where about a quarter of all our *Enterobacter cloacae* are de-repressed mutants. Bearing in mind that these mutants only appear to survive if the antibiotic pressure is so great that the bacterium has to make this enormous concession just to survive, it is not surprising that less drastic resistance mechanisms can survive far more readily.

Until the emergence of de-repressed mutants of *Enterobacter cloacae*, cephalosporin use was supremely successful. The so-called second-generation cephalosporins such as cefuroxime and the cephamycins, such as cefoxitin, became the mainstay of hospital infection control. These cephalosporins were not resistant to all β-lactamases and they were not very effective against the non-fermenting bacteria such as *Pseudomonas aeruginosa* and *Acinetobacter baumannii*. The later-generation cephalosporins were developed to overcome these deficiencies. These were made by the addition of an oxime group at the 7-position of 7-amino-cephalosporanic acid (7-ACA). This cephalosporin group is exemplified by 1) ceftazidime with a carboxylic alkoxyimino group, resulting in slow outer membrane penetration in the Enterobacteriaceae; thus although this cephalosporin confers a significant degree of anti-pseudomonas activity it is less effective against species such as *Klebsiella*, and 2) cefotaxime with a methoxyimino side-group, which gives the molecule faster penetration, giving a good anti-*Klebsiella* profile but far less anti-*Pseudomonas* activity. *Klebsiella* species had been successfully controlled by the previous generation of cephalosporins and it was assumed that they would continue to pose no problem for therapy.

The most successful plasmid-encoded β-lactamases in *Klebsiella* species were TEM-1, the close relative but less successful TEM-2 and the distant cousin SHV-1. Between them these β-lactamases accounted for considerably more than 90% of all transferable β-lactamase-mediated resistance. The genes encoding these enzymes are all located on transposons and well able to migrate into other species. By the time the third-generation cephalosporins had been introduced, the genes for TEM-1 and TEM-2 had extensively invaded this species. SHV-1 is thought to have orginated from *Klebsiella* in the first place. It is uncertain whether it was the fact that these resistance genes had migrated into *Klebsiella* species by the time that the third-generation cephalosporins were introduced or whether the 7-oxime cephalosporins provided an easier target for resistance to develop; however, the most rapid emergence of resistance coincided with the introduction of these drugs.

The elimination of *Escherichia coli* as a feared pathogen in the Intensive Care Unit was the result of treatment with cephalosporins; almost any cephalosporin was effective. This appears to have led to complacency with the view that any cephalosporin should be able to control *Klebsiella* as well. Problems in the past, particularly with the emergence of the gentamicin-resistant *Klebsiella* in British hospitals in the 1970s, suggest that this bacterial genus would not submit as easily as *Escherichia coli*. The feared pathogen was *Pseudomonas aeruginosa* as it was perceived to be the gram-negative bacterium that was most likely to become multi-resistant. In fact when the 7-oxime, third-generation cephalosporins were introduced in the early 1980s, *Pseudomonas aeruginosa* was already declining as a hospital-acquired pathogen. However, many clinicians thought, perhaps erroneously on many occasions, that the therapy that they administered should ensure that sufficient cover was provided for *Pseudomonas aeruginosa*, despite the fact that it would only be found on a fraction of occasions compared with species of the *Klebsiella* genus.

This meant the favouring of those cephalosporins that were more active against the non-fermentors, but they were less effective against *Klebsiella*; they penetrated more slowly and were less efficient at killing these bacteria. An enormous proportion of the clinical *Klebsiella* species had already had been invaded by the transferable β-lactamases and they were continually challenged with sub-optimal concentrations of inappropriate cephalosporins. It should not be surprising in retrospect that this would lead to resistance; the only real surprise was how the resistance was achieved. The mutation of the TEM and SHV β-lactamases was totally unexpected and considered to be impossible. The extended-spectrum β-lactamases resulted from mutations in the active site of either TEM or SHV enzymes. Usually the first mutation was of little clinical importance but the second mutation could be absolutely devastating, providing the host bacterium with complete protection against a whole class of cephalosporins. Interestingly, the extend-spectrum mutations never gave comprehensive resistance to the second-generation cephalosporins and this might explain why they did not emerge while these drugs were in the ascendancy. The mutation events almost always occurred in the *Klebsiella* genus, usually *pneumoniae* and *oxytoca* species, and, as they occurred in genes located on plasmids, they then spread into *Escherichia coli* and other members of the Enterobacteriaceae. The diversity of the mutations makes it very difficult to determine how they spread and exactly how they developed. We can only speculate at this; however, once a certain combination of mutations had occurred, clonal outbreaks appeared. This was seen early in the emergence of the extended-spectrum β-lactamases; there were outbreaks of *Klebsiella* strains harbouring SHV-2, particularly in hospitals in France and Germany during the early 1980s. SHV-2 is very similar to the parent enzyme except that it has a serine group substituted at position 238 instead of the normal glycine. This provides extensive resistance to almost all cephalosporins, both the fast and slow-penetrating ones. The success of this β-lactamase and its host *Klebsiella* probably resulted from extensive use of cefotaxime or similar cephalosporin. In a recent outbreak in Aberdeen during the 1990s, there has been an outbreak of a multi-resistant *Klebiella pneumoniae* carrying an SHV-2 β-lactamase and this has been associated with the clinical use of cefotaxime; thus closely related strains have been isolated on over 450 occasions in a small georgraphic area. A further mutation at position 205, changing

222

Magic bullets

arginine to leucine, can occur to give SHV-5. Biochemical analysis of the β-lactamase does not show what advantage this might have over SHV-2 but the techniques used may be insufficient to demonstrate the subtlety of the difference. The fact that it has evolved suggests that this second mutation is important. It is certainly prevalent in *Klebsiella* isolates responsible for hospital epidemics throughout the world. There have been outbreaks in Germany and particularly in Greece and Turkey. It is possible that the first mutation resulted from extensive use of a rapid-penetrating cephalosporin such as cefotaxime and that the second mutation came after a switch to a second cephalosporin. As it is impossible, biochemically, to distinguish SHV-5 from SHV-2, it is difficult to identify which cephalosporin or even which type of cephalosporin was responsible for the selection of this mutation. Nevertheless, of the SHV-derived extended-spectrum β-lactamases, SHV-2 and particularly SHV-5 appear to be the epidemic genes.

There is a vast array of TEM-derived extended-spectrum β-lactamases that have emerged largely in *Klebsiella* species. However, only a few are found with any regularity. In London, at the start of the extended-spectrum β-lactamase outbreak, there was a predominance of TEM-10. This is derived from TEM-1 and has a double mutation. The first was almost certainly the alteration of arginine at 164 to serine. This gives a very low level of resistance, which is usually not clinically significant. However, it only gives any degree of insusceptibility to the slow-penetrating cephalosporins and none to fast-acting drugs. This means that it would be virtually impossible for fast-penetrating cephalopsorins to select this mutation. The second mutation in which glutamic acid at 240 is substituted by lysine pulls the β-sheet away from the active site; this promotes the binding of the slow-penetrating cephalosporins, particularly ceftazidime, rather than cefotaxime. It gives no clinical resistance to cefotaxime at all and thus these two mutations have probably resulted from the use of the slow-penetrating cephalosporins alone. In France the situation was different. The predominant β-lactamase was TEM-3. This is a TEM-2 derived enzyme and has undergone at least two mutations. As the figure overleaf shows, the stages at which these mutations occurred may not be straightforward. *Klebsiella* species carrying the TEM-3 β-lactamase became epidemic in France and were found in hospitals hundreds of miles apart. It did appear that this β-lactamase/*Klebsiella* combination had an advantage that the others lacked. The TEM-3 β-lactamase confers high levels of resistance to the fast-penetrating cephalosporins, though the diagram suggests that the early mutation might have been a slow penetrator and then, when this began to fail, a switch was made either to a penicillin and β-lactamase inhibitor or a fast-penetrating cephalosporin or both. We do not know enough about the host strain to state whether this was a significant factor.

Ferdinand Baquero in Spain has been intrigued that so many of the extended-spectrum β-lactamases are derived from TEM-2, which is surprising

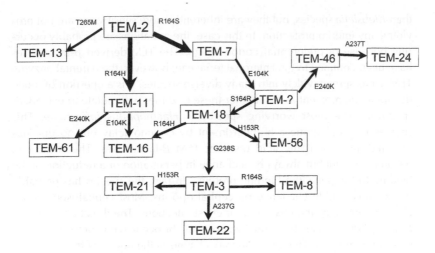

Selection of the extended-spectrum TEM-2 β-lactamases

considering how rare it is compared with TEM-1. He speculates that TEM-2 provides greater survivability to cefotaxime. It does not provide resistance but rather confers the capability to die less quickly on cefotaxime challenge. If TEM-2-containing strains survived challenge for longer than TEM-1, there would be more opportunity for mutation to occur. It might also be expected that there would be a rise in the carriage rate of TEM-2 in areas where cefotaxime is widely used but this has not been reported yet. Nevertheless, this may explain the success of this enzyme as a progenitor.

Perhaps the most interesting phenomenon of the extended-spectrum β-lactamases is that they almost always emerge in *Klebsiella* species. This genus must provide the optimum environment for mutations to occur because, once they have occurred, they can be transferred by their host plasmids into other species of bacteria and can survive well in them. These mutations rarely occur in these species. It may be that *Klebsiella* species do have an inherent level of resistance to β-lactam antibiotics; it is high against the penicillins and, although lower against the cephalosporins, does provide some degree of protection against these antibiotics. This view would be supported by the emergence of other resistance genes in this genus. An alternative view would be that *Klebsiella* species permit the mutations more readily than other species. All bacteria have favoured codons for each of the amino acids and those used by *Klebsiella* may be those that are readily formed by the mutations in the TEM and SHV β-lactamases.

A similar phenomenon is the emergence of mutations in the TEM β-lactamase that give clavulanic acid resistance; however, these almost never occur in *Klebsiella* strains and are virtually exclusive to *Escherichia coli*. Certainly *Escherichia coli* is challenged more regularly with co-amoxiclav than *Klebsiella* species, but they are inherently more sensitive so are not providing any interim protection. In this case, the mutations are probably occurring because of preferential codon usage. The TEM-derived β-lactamases have been confined to the Enterobacteriaceae; however, the parental enzyme TEM-1 has spread widely into many diverse species. It is a question of much debate at the moment as to whether these enzymes will mutate to extended-spectrum. The most worrying species is *Haemophilus influenzae*. This species has undergone a bombardment by co-amoxiclav but no one has reported any clavulanic acid resistant TEM β-lactamases. The strain can become resistant but always by a change in penetration or a reduction of the binding to the penicillin binding proteins. In fact, the species has probably had relatively little challenge by cephalosporins. Most cephalosporins are given parenterally and this is not for chest infections. The direct challenge of *Haemophilus* species by cephalosporins might occur with some of the few cases of meningitis. This situation may change as the nuclei of the third-generation cephalosporins are altered to allow oral use. If these drugs are used extensively in the community, the opportunities for mutation to occur and be selected may increase considerably. We are actually seeing a rise in cephalosporin resistance in Scotland and this may reflect the increased usage of oral variants; however, none of these resistant strains has a modified TEM enzyme. Interestingly, the TEM-1 β-lactamase in *Haemophilus influenzae* is slightly different from that in the Enterobacteriaceae. There are two different amino acids at the outer edge of the active site. As there is no alteration in the biochemical properties of the TEM-1 β-lactamase in *Haemophilus influenzae*, these alterations are not thought to be significant. They may, however, reduce the ability of the molecule to undergo mutation to extended-spectrum activity. It is not known how *Haemophilus* species acquired the TEM-1 β-lactamase in the first place; it would be unlikely to have obtained it from *Escherichia coli* directly, as these bacteria are rarely in the same micro environment. However, *Klebsiella pneumoniae* is a frequent infectious agent in the chest and it may often be found alongside *Haemophilus influenzae*. *Klebsiella pnuemoniae* responsible for chest infections have been known to carry the SHV-5 β-lactamase. It is thus highly likely that this species could carry an extended-spectrum β-lactamase into the chest and a simple transfer would place it in *Haemophilus influenzae*. The selective environment of oral cephalosporins would do the rest.

PSEUDOMONAS

Pseudomonas aeruginosa has always been considered a potential pathogen and may be an early example of an environmental organism entering the clinical setting. It has always been considered to be an opportunistic pathogen that would infect the patient who was compromised in some manner and whose defences were reduced. When the main nosocomial pathogens were members of the Enterobacteriaceae, *Pseudomonas aeruginosa* was considered to be the cause of the most problematic infections. The bacterium was multi-resistant and did not respond to the antibiotics to which other invaders of Intensive Care Units responded. Indeed, special antibiotics had to be developed to deal with *Pseudomonas aeruginosa* alone; these started with carbenicillin followed by azlocillin, mezlocillin and some of the cephlaosporins.

   Pseudomonas aeruginosa has a number of virulence mechanisms associated with it; however, in reality these were probably not to improve the spread of this genus through the animal population but rather to provide some protection in the environment. It has a reduced number of porin channels so it can limit the number of small hydrophilic molecules, including antibiotics, taken up into the cell. The environmental origins of *Pseudomonas aeruginosa* are probably the reason why it has had to develop this general mechanism of reduced penetration. Although this resistance may not be the most efficient defence against antibiotics and even if it does not allow *Pseudomonas aeruginosa* to grow, it provides this species with the capability to survive longer in an antibiotic-containing environment than most of its competitors and thus provides the "breathing space" to develop or import more effective resistance genes. The species has also been considered to be multi-resistant because it may contain a *mar* gene or genes, which would encode an efflux pump that would export small molecules out of the cell. This would also be a general defence mechanism learnt while the strain was in the soil, protecting itself against the invasion of antibiotics released by its competitors.

   In recent years, *Pseudomonas aeruginosa* has largely been controllable by ceftazidime, fluoronated-quinolones and imipenem but the development of resistance to the cephalosporins by the de-repression and hyperproduction of the class C β-lactamase and the emergence of mutated *gyrA* genes and fluoroquinolone impermeability mutants suggest that the power of these drugs will not always persist. The latest concern is the description of plasmid-encoded imipenem resistance mediated by the class C β-lactamase IMP-1 found in some Japanese strains. There are many, many more imipenem-resistant *Pseudomonas aeruginosa* that do not have this metallo-β-lactamase than do and there is no question that *Pseudomonas aeruginosa* will become increasingly resistant to carbapenems; the doubt is whether it will be mediated by a relatively rare plasmid-encoded β-lactamase. The predominant mechanism of resistance appears, like cephalosporin resistance, to be a combination of hyperproduction of a class C β-lactamase and reduced penetration. If class C β-lactamases are the harbingers of carbapenem resistance, inappropriate cephalosporin use can select them as readily as a carbapenem alone.

   This raises serious questions as to how long this organism remains vanquished. It is a problem for which we might not need an answer. It is a surprising fact that *Pseudomonas aeruginosa* has become an increasingly less significant nosocomial pathogen. Whether this is because it does not favour infection in the immunosuppressed as readily as some other multi-resistant pathogens or whether the antibiotic cocktail we now give does not favour its survival, despite the emergence of resistance, is not clear. The good news is that it is not high in the list of untreatable pathogens and clinicians who still dictate their therapy to guard against invasion by *Pseudomonas aeruginosa* might consider whether their antibiotic policies are, in reality, leading to the replacement of *Pseudomonas aeruginosa* with bacteria that are far more difficult to control.

Pseudomonas aeruginosa is also a predominant cause of respiratory infection in children with cystic fibrosis. This genetically inherited disease and subsequent infection with *Pseudomonas aeruginosa* was usually a death sentence before the age of 20. Now we have learnt a great deal about the pathogenicity of the bacterium and more importantly the correct antibiotic and physiotherapy management of the respiratory infection. Sufferers from the disease regularly survive to 40 and beyond and it is soon expected that those who might be exposed to *Pseudomonas aeruginosa* infections alone could be successfully managed to live a nearly full life span.

Unfortunately, this optimism does not exist for patients infected by *Burkholderia cepacia*. This species was considered to be a member of the *Pseudomonas* genus. It had a similar appearance on some laboratory media and had similar biochemical reactions in early tests. We now know that genetically the bacterium is quite distinct. It is also an environmental bacterium and does share the predisposition to multi-antibiotic resistance. When this organism invades the lungs of a cystic fibrosis sufferer, the prognosis is bleak. Apart from the heightened pathogenicity of this bacterium, its multi-resistance makes it virtually impossible to eradicate. It is already resistant to most antibiotics that can eradicate *Pseudomonas aeruginosa* and it can acquire resistance to imipenem quite readily. A species-specific metallo-β-lactamase has been associated with *Burkholderia cepacia* and it may be able to acquire further imipenem resistance genes. Genetic anlysis of the bacterium shows that it is able to mutate rapidly and it would be expected to develop resistance to any antibiotic that was directed against it. The bleak outlook found in cystic fibrosis may soon also be found in other hospitalised patients as this pathogen infects the immunosuppressed in the Intensive Care Unit. Its major selective characteristic may be its multi-resistance. If that is the case, we may see this bacterium invade the immunosuppressed.

ACINETOBACTER

The emergence of cephalosporin resistance in the Enterobacteriaceae has driven the use of the fluoro-quinolones and carbapenems. This regime may have also stemmed the invasion of *Pseudomonas aeruginosa*. The dismissal of the traditional pathogens left a vacuum which has partially been filled by the genus *Acinetobacter*. Twenty years ago this environmental organism was considered entirely non-pathogenic and an oddity when found in hospitals. It was certainly no threat to Intensive Care Units. Unfortunately, this view of *Acinetobacter* still persists with some prescribers and insufficient care is taken to control it. It is probably now the most important cause of hospital-acquired gram-negative infection that we currently have to deal with and we are clearly rapidly losing this struggle.

Acinetobacter species appear to have been very successful environmental survivors; during their struggle for nutrients they will have been exposed to a myriad of antibiotics, albeit at low levels, for far longer than most true clinical bacteria. In their original environment, the soil, they were forced to evolve defence mechanisms to various antibiotics synthesised by their aggressive neighbouring bacteria which produced antibiotics to promote their own survival. Thus they have the potential to develop new resistance mechanisms very rapidly, "learnt" from aeons of exposure to successive environmental antibiotic attacks. The early clinical *Acinetobacter* strains, isolated 20 years ago, were not highly resistant to the clinical antibiotics when they first emerged as a clinical problem. It is likely that they were already resistant to many of the antibiotics abundant in the soil but are not used clinically because of toxicity, although this is unlikely ever to be known. They do, however, appear to have developed the machinery for rapid acquisition of resistance genes and when they were exposed to the clinical antibiotics, resistance followed very rapidly.

The early *Acinetobacter* strains responded to treatment with the tetracyclines, aminoglycosides, nalidixic acid, and the penicillins. The speed at which *Acinetobacter* could adapt to confer resistance was awesome as within three years, most clinical isolates had become resistant to most antibiotics commonly used at that time. As the bacterium was not considered as an important pathogen, this rapid learning curve of *Acinetobacter* was probably not the result of therapy against the bacterium but inadvertent challenge from the treatment of other pathogens. The pathogenicity of *Acinetobacter* does not appear to have changed and it is able to invade more readily because of the procedures that we now use. This bacterium seems to prosper better than most of the others because of its propensity to develop antibiotic resistance; the very act of rapid evolution of resistance genes is the strain's major pathogenic attribute. We now face problems with infection by this bacterium merely from our use of antibiotics alone. The more resistance genes it acquires, the more pathogenic it becomes. There is now almost universal resistance to many of the older antibiotics, including most penicillins and first- and second-generation cephalosporins, most aminoglycosides, chloramphenicol and tetracycline.

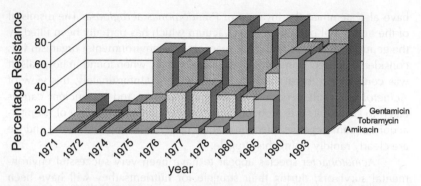

Data from Bergogne-Berezin (1996). Acinetobacter, CRC Press, p133.

Progressive rise in Aminoglycoside resistance in *Acinetobacter* spp

The acquisition of cephalosporin resistance was probably the turning point for *Acinetobacter* species. It was able to acquire resistance to these β-lactams far more quickly that *Klebsiella* or the other Enterobacteriaceae. Unlike *Klebsiella* species, the *Acinetobacter* genus acquired universal resistance to second-generation cephalosporins long before the third-generation cephalosporins were introduced, and there was no significant resistance before the middle of the 1970s, when only the first-generation cephalosporins were used. The acquisition of multi-drug resistance in *Acinetobacter* species coincided with the introduction of second-generation cephalosporins; that is not to state that they were solely responsible, but their use may have selected *Acinetobacter baumannii* variants that were particularly effective at acquiring new resistance genes, rather like the MRSA. During the 1970s, these bacteria were becoming an increasing problem but were still not as important as *Escherichia coli, Klebsiella* species or *Pseudomonas aeruginosa*. In the early 1980s, *Klebsiella* species were acquiring resistance to cephalosporins, through extended-spectrum β-lactamases, which were spreading rapidly in *Escherichia coli* and other Enterobacteriaceae. Therefore, an urgent switch in therapy was needed to control *Klebsiella* strains and this coincided with the launch of the fluoroquinolones in the mid-1980s.

The first fluorinated quinolone was perfloxacin, perhaps the weakest member of the group. This did prove successful in controlling *Klebsiella* and had the effect of reducing its importance as a pathogen of the Intensive Care Unit; however, perfloxacin use created a vacuum which would be filled by a bacterium that had a talent for multi-drug resistance. In the mid-1980s, the majority of the *Acinetobacter* strains in France were sensitive to fluorinated quinolones, but by 1989, 75% were completely resistant. For the most part, if a strain is resistant to one fluorinated quinolone it becomes resistant to the whole class and so the French *Acinetobacter* strains were resistant to the more powerful drugs that were yet to be launched. In the United Kingdom, the problem of cephalosporin-resistant *Klebsiella* was not acute and thus the need for fluorinated quinolones was not so immediate. There was no requirement to introduce perfloxacin and, in fact, it has never been licensed in this country. Ciprofloxacin was introduced instead. It is much more powerful then perfloxacin and may have been crucial to the delayed emergence of fluoroquinolone resistance in *Acinetobacter baumannii* in the United Kingdom. This is further support for the general rule of chemotherapy, to use the most powerful member of a drug class, ciprofloxacin in this case, to reduce the risk of selecting resistant variants. However, such a strategy was not infallible because in Edinburgh, in 1994, almost all the *Acinetobacter* spp were fluoroquinolone sensitive, but after two outbreaks, 68% of the *Acinetobacter* strains became resistant.

The only antibiotics left that are powerful enough against *Acinetobacter* species are the carbapenems, so imipenem has been used extensively to treat infections caused by *Acinetobacter* infections but this has already led to significant levels of carbapenem resistance. In the United Kingdom, the incidence of carbapenem resistance is still virtually non-existent and this probably results from a reluctance to use imipenem unless we really need to. This cautious approach has served us well, though we may not be able to continue this for long. There have been problems with carbapenem resistance in *Acinetobacter* strains isolated in this country. The first plasmid-encoded carbapenem resistance in Europe was found in an *Acinetobacter* strain isolated in Edinburgh though it does not appear to be spreading at the moment. Even so, it means that we shall have to use carbapenems very cautiously. Some countries use the β-lactamase inhibitor sulbactam to control *Acinetobacter baumannii* infections. We currently do not use sulbactam; it is surprisingly not available in the United Kingdom, but if we get into trouble with this organism, we still have the option to use sulbactam though this will be a fairly desperate solution.

Buenos Aires, Argentina, has up to 35% of *Acinetobacter baumannii* isolates which are carbapenem resistant and, unlike their British counterparts, these strains are also resistant to sulbactam. Sulbactam has been needed in Argentina to cure *Acinetobacter* infections and resistance is now around 30%. This is the doomsday scenario, an important pathogen with no viable antibiotic control. *Acinetobacter baumannii* is a frequent invader of Intensive Care patients and the inability to treat them is a disaster. All we can do is apply old barrier nursing techniques and try to prevent cross-infection; however, this organism specialises in the infection of the immuno-compromised and the risk of infection with this totally resistant bacterium is likely to prevent the immunosuppression required for transplantation, for where imipenem resistance is prevalent, transplantation may well have to cease.

By some strange coincidence, I have written some of this chapter while I am attending the ANKEM conference in Antalya in southern Anatolia. This meeting is discussing the incidence of resistance in Turkey and there is a paper which declares that resistance levels to imipenem are around 89%. Without verification, I cannot state whether this is true but, if it is, there is a real problem and the risk of infection after immunosuppression of the patient is so great that I suspect that they will have to suspend such operations until they reduce the risk of infection. The economic effects of invasion by multi-resistant strains was demonstrated in a hospital in France. The emergence of carbapenem-resistant *Acinetobacter baumannii* was problematic in France, where patients in two separated ITUs became infected with two different strains. Epidemiological studies demonstrated that environmental contamination had been the reservoir and that patients ran the risk of infection as soon as they entered the Intensive Care Units. Control was only achieved after the extremely expensive exercise of complete closure of both ITUs and thorough decontamination.

MULTI-DRUG-RESISTANT BACTERIA

It is possible that the threat of carbapenem resistance in *Acinetobacter baumannii* does not become a major problem in some hospitals. In the United Kingdom, they are still very rare but we are usually cautious about antibiotic usage. This does not mean that we are free of resistance problems but rather that they might manifest themselves in a different manner. The use of fluoroquinolones reduced the importance of *Klebsiella* in many hospitals; they were not replaced by fluoroquinolone-resistant *Klebsiella* but rather by the ascendancy of *Acinetobacter*. This may also be true for the treatment of *Acinetobacter* species with carbapenems, as in the United Kingdom we have seen a rise in the emergence of a multi-resistant bacterium, *Stenotrophomonas maltophilia*. This bacterium, first classified as *Pseudomonas maltophilia* and then *Xanthomonas maltophilia*, is a non-fermenting bacterium like both *Pseudomonas aeruginosa* and *Acinetobacter baumannii*; but unlike either it is inherently resistant to virtually all antibiotics. It is a poorly studied organism because it was not considered pathogenic, merely a coloniser. Its multi-resistance allows it to fill the vacuum left by more virulent pathogens. This species is inherently resistant to cephalosporins, fluoroquinolones and carbapenems. It shows some sensitivity to co-trimoxazole, but the use of this combination is a fairly desperate measure for the control of *Stenotrophomonas maltophilia*.

HOW DOES A MULTI-RESISTANT BACTERIUM SPREAD?

In the modern antibiotic era, multi-resistance must be considered as a pathogenic feature as potent as the ability to attach to the vulnerable site of infection, as many of the bacteria that cause gastrointestinal infections do in the small intestine, or prevent the body's defences eradicating the infection, as in the case of many *Staphylococcus aureus* infections. In the case of *Stenotrophomonas maltophilia*, multiple resistance to antibiotics is probably the only pathogenic feature. The patients that are infected are almost always immunosuppressed and would be susceptible to infection by many bacteria, however harmless to the healthy individual. It is likely that most of the infection is spread by the staff and by the transfer of patients through the health-care system.

We have already considered the failure to wash hands as a means of spreading infection; since Semmelweis first demonstrated its importance 150 years ago in Budapest, this simple precaution has been ignored. It does not matter whether the hospital is in a poor developing country or a high-technology unit in North America, breaches of this simple precaution are common and it is probably the single most common cause for the spread of infection, so if mentioning it again seems like repetition, I make no apology because it is so important that it should be repeated constantly. Patients should refuse to be treated by health-care workers that have not washed their hands after attending to the previous patient. There is no excuse for transmission of infection by this route. Hospitals must ensure that there are sufficient wash-hand basins in their wards and particularly in their Intensive Care Units. In the latter, there should be one close to each bed. Hospital designers rarely take the spread of infection seriously if indeed they really know anything about it. The design of many new hospitals is driven by the need to save money and usually this conflicts with the control of infection.

The proper sterilisation of instruments is also essential to control the spread of infection. This sounds so obvious that it hardly would seem to require a mention. Sterilisation means the complete removal of all bacteria and other microorganisms; with those items required for the incisions and traditional procedures in operations this is usually easy to achieve; they are simply autoclaved which effectively means raising them to a temperature of 133 °C for more than three minutes in saturated steam. This is virtually guaranteed to kill all infectious agents except the prions that cause BSE, but there is currently no risk that these prions are a significant cause of infection in hospital. The nature of operations is now changing; the so-called key-hole surgery does not call for traditional instruments. It requires a minute video camera and a miniature lighting system. These cannot be sterilised by the traditional steam heat methods, as these procedures would destroy the intricate circuitry. Instead they have to be sterilised by a mixture of chemicals and mild heat treatment, which really cannot guarantee to remove all microorganisms, but merely to reduce their numbers to an acceptable level. The problem for many of our modern patients is that there is often no minimum level of bacteria that would be considered safe if their immune systems are not functioning.

The second problem is the spread of resistant bacteria within the hospital. Infections that originate in hospital are usually known as nosocomial infections and have traditionally been associated with the transfer of bacteria from one patient to another or from the hospital environment to the patient. In the past, the patient would go to hospital perhaps to have an operation. After the operation, the patient would be moved to a ward where he or she would stay until almost fully recovered. The treatment of any infection was usually performed with parenteral antibiotics, which necessitated a stay until therapy was completed as nursing staff were required to administer the drugs by injection. In modern medical practice, the pressure on beds is greatly increased and the patients are moved regularly to wards with decreasing dependency. The patients, while they are severely ill, are given parenteral antibiotics but, as soon as they are able, they are given oral versions of the same drug so that they may be released home. The patients may also be sent to convalescence facilities, rehabilitation centres or a nursing home; these are part of the modern health-care network and, for infection, may be considered as part of the hospital. The movement of patients spreads infection. There are now many reports of patients in nursing homes with MRSA and some homes would like to have patients screened before they are admitted. Similarly, hospitals within the same area may move patients to specialised facilities in their health district, further opportunity for infection to be transmitted. We now have to consider the whole complex of health-care institutions as part of the hospital complex and, with this information, consider how to control the spread of infection and the transmission of multi-resistant bacteria. This is not going to be easy as the successful control of hospital infection invariably is not commensurate with budget reductions.

10

The future of antibiotics – has resistance switched out the light at the end of the tunnel?

In 1969 the Surgeon General of the United States declared that we can close the book on infectious diseases. He believed that antibiotics had tamed all bacterial infections for all time. Just 20 years later, in 1994, the then Chief Medical Officer of Scotland predicted that we will have run out of antibiotics by 2020; the rise in resistance would far outstrip the dwindling supply of new antibiotics. This was in response to two articles that appeared in *Time Magazine* and *Newsweek*, which predicted the end of the Antibiotic Era. Indeed, the House of Lords reported in 1998, in their Select Committee on Antibiotics and Resistance, that we must urgently engage in more, much-needed research to find new antibiotics and devise strategies to overcome resistance, so far from closing the book, we must examine how we have gone from a period of immense optimism about the power of antibiotics to a general consensus of doom.

The reasons for this change have been covered in many of the chapters of this book but reviewing the causes of the rise in resistance may not provide the solution to the future of the control of infectious diseases. The period from the end of the Second World War to the start of the 1960s was the golden age of antibiotics. In 1961, nalidixic acid was discovered and, although this was the first quinolone, it was the last completely new chemical structure that was to be introduced into the antibacterial armentarium. There have, of course, been many new antibiotics launched since then but all of them have been modifications of previously used chemical nuclei. This means that resistance to the modified drugs had already been learnt either by bacteria or by their plasmids and other mobile genetic elements. The resistance that had emerged to previous members of that class of antibiotics would either provide cross-resistance to the later versions or provide the progenitors for new resistance mechanisms to emerge. Certainly, the development of the fluoroquinolones was a significant improvement over the early members of the class, such as nalidixic acid; however, widespread use of nalidixic acid creates resistance problems for the whole class of quinolones. We face a crisis that has a multifactorial cause and unless we face up to it, we will provide our children with limited or no capability to deal with the bacterial infections, particularly those acquired in hospital, that our grandparents would have understood and feared. The Antibiotic era will be seen as that brief period of time when we were able to cure infectious diseases.

If we examine the strategies and attitudes of the past 30 years we can begin to understand why this situation occurred. Antibiotics have been accepted by the medical profession with more enthusiasm than any other group of pharmaceuticals; they have always been used without the consideration that we would ever have to preserve the drugs that we currently possess because it has always been assumed that new drugs would always be introduced to save us from the resistance that has emerged to the previous generation. This is a general but naïve view because no new classes of drugs have been introduced for nearly 40 years; rather, more powerful versions of the antibiotics that we already possess are launched with great publicity and an extensive promotion campaign. Sometimes these drugs are heralded as major breakthroughs; at least two were presented as such on the *Tomorrow's World*. However, they were simply variations on previous antibiotics and resistance to these early drugs has already started the development of resistance to new wonder-drugs. This often results in a reduction in the time that resistance takes to emerge and as the later versions of the various classes of antibiotics are introduced the speed with which resistance appears is alarmingly quick.

The second obstacle for the development of new antibiotics is that our expectation of their safety is much greater. We have seen how the licensing authorities now expect safety profiles that would allow almost no antibiotics to be given for simple infections, based presumably on the premise that if they are licensed, there will be someone, somewhere who will use them for a simple infection despite the fact that they were designed for severe hospital infection.

The third disincentive is that there may be insufficient money from the market at which a particular antibiotic is targeted. The major markets for antibiotics are in the developed world and resistance is mainly in nosocomial infections, so this is where new and inevitably expensive antibiotics will be used. There may be insufficient usage to pay back the development costs, whereas if the drug is used for community infections, it may be prescribed regularly by every community practitioner, perhaps once or twice every surgery session. The return on the investment could be enormous.

These are significant disincentives to the pharmaceutical companies to produce new antibiotics. The economics of antibiotic discovery and development was recently demonstrated in a recent talk by Alan Rauch from Glaxo-Wellcome. Although pharmaceutical companies are amongst the most successful of all commercial companies, they have to invest 18.5% of their sales in research and development. No other industry has to invest anything like this figure; the electronics industry invests just 5% and it is the second-largest investor in R&D. Such an enormous commitment to R&D makes the pharmaceutical industry particularly vulnerable to mistakes in strategy; if they make significant investments in two or three compounds that fail to reach the marketplace, this can bankrupt even the most robust company. Many financial observers will have noticed that all the major pharmaceutical companies have striven in the past few years to make themselves international and large enough to ride through the lean years. The majority of companies are now the result of multinational amalgamations or takeovers. A figure of 30,000 personnel has been suggested as the minimum requirement for a viable company.

The whole process of discovering new drugs and then bringing them to market has lengthened considerably in time and the costs have been subject to an inflation rate that would ruin some governments. The average cost in 1976 to bring a new drug to market was $54 million and the whole process took just 7–10 years. Just 10 years later, in 1986, the costs had risen to $125 million and the time taken was 10–12 years. By the beginning of the

Pre-clinical:	The earliest phase of development that primarily examines the compound, obtained by small scale synthesis, for laboratory efficacy and safety.
Phase I:	Evaluation of the safety and tolerability of the compound, invariably performed on volunteers under stringently controlled conditions.
Phase II:	Initial evaluation of efficacy in a small number of patients and establishment of optimum dosing.
Phase III:	International clinical trials with large number of patients to establish clinical and pharmacological benefit. Also used to identify potential safety problems.
Submission:	Documentation of the above sbmitted to regulatory authorities for licensing application.
Launch:	Commercial release of drug, sometimes under strict guidelines from regulatory authority.
Phase IV:	Post launch surveillance to determine whether predicted safety profile is valid, to identify further problems and improve drug usage.

Phases of drug development

next decade (1990), the costs had more than doubled again to $359 million and development times had risen to 12–15 years. The latest estimates for 1996 are that the estimated cost is $500 million and the development will take 15 years. Under special circumstances, such as urgent medical need, the development time can be reduced; however, this will only occur in cases of obvious medical need. When the anti-HIV drug zidovudine underwent clinical trials it was shown to have dramatic therapeutic effects and so it became available earlier than would be predicted, entirely because of its perceived immediate benefit. This drug does have a significant number of side-effects but the lack of any, let alone safer, alternatives drove the fast-tracking of this drug to market. Such acceleration is very unlikely to occur for an anti-bacterial drug, particularly one based on a novel chemical nucleus. The US Food and Drug Administration still cite the case of thalidomide as a reason to remain vigilant and not allow drugs to be launched until their profiles are thoroughly tested.

The discovery of a new chemical for therapeutic use might take between 2 and 10 years. It is likely to come from a major research programme comprising a multidisciplinary research team. In some companies, they have set up rival research teams and presumably reward the successful and disband the unsuccessful. Probably 10 years is the very maximum time that a company might invest in a discovery project if there are no obvious rewards, and many would cease long before that. New, prospective anti-bacterial drugs would immediately undergo pre-clinical testing. This would mainly comprise bacteriological laboratory tests, to determine the inhibitory concentration against a range of standard pathogenic bacteria. Is the drug stable, and what effect does temperature or pH have on the drug composition? This process occurs with hundreds of prospective drugs concurrently and it is essentially a screening process. Only drugs that show no obvious disadvantages pass through to the next stage. Approximately one drug in 500 makes it through this hurdle and it is relatively cheap to expel a drug at this stage.

The first tests will be in animals, which have usually involved at least two animal groups. Rodents are always perceived as the classic laboratory animal; however, for antibiotic experiments they are often quite unsuitable because they have completely different pharmaco-kinetics from humans for many drugs. Thus one of the test animals should be a species other than a rodent. Phase I of the drug development is the first introduction into humans, always healthy volunteers. This is primarily to determine if there any unforeseen side-effects. The study group go to the test centre to have the drug administered under close medical supervision. These volunteers are investigated thoroughly for obvious effects and various biochemical examinations are made to see if the drug alters any of the body's enzymic functions, particularly alterations in liver and kidney enzymes; there are sometimes quite unexpected and unusual side-effects. This stage may take approximately a year.

If the drug passes Phase I, Phase II is evaluation of the drug for a clinical indication. This will determine what type of infections will be most suitable for treatment with this new antibiotic. This will require evaluation of the pharmaco-kinetics and the capability of the antibiotic to effect a cure. In particular, this phase will be used to find the optimum dose and the dosing intervals. About 20% of all compounds that enter this phase make it to final clinical launch. At the end of approximately two years, the antibiotic should be ready to enter full-scale clinical trials.

Phase III is the testing of the drug under clinical conditions. This involves the inclusion of thousands of patients and the antibiotic is released specifically for named patients. This usually requires direct patient consent and this is expected to be informed where the patient has had all the benefits and risks described, where this is possible. The main purpose of this stage is to determine whether the new antibiotic actually works and whether it has a significant activity when compared with another antibiotic or placebo. Placebo trials are the classic method for drug trials but if the patients are severely ill, placebo trials could be considered unethical; those on the placebo may be placed at risk because they are not being treated. Many trials now compare antibiotics to other drugs that might be used to treat similar infections. In these cases, the onus is not so much to demonstrate that it is much better than the comparator, as it would be if it was tested against a placebo, but rather to demonstrate that the new antibiotic is no less effective than the comparator. The early tests in Phases I and II have been performed on relatively few volunteers and if there are rare side-effects, these may not have been revealed. It is expected that the thousands of patients entered at this stage should demonstrate if there are likely to be any significant problems. It does not cover every eventuality; fairly recently, some fluoroquinolones had passed through this stage before it was found that there were problems with phototoxicity, a discovery which forced immediate withdrawal. Phase III can take approximately four years, and it is also used by the pharmaceutical company to prepare the clinical community for a new antibiotic. On discovery, the drug was given a working number, usually prefixed with an acronym of the company's initials; at this stage it is given both a generic and often a marketing name. The latter is one that the company will use in its literature and often enormous effort is invested in producing a name that is easy to remember and prescribe.

As Phase III is completed, the company will prepare for the licensing of the drug. The major international hurdle that it has to pass is the US Food and Drug Administration. The standards are very high and all the data of the three previous phases are included. There are an increasing number of tests that have to be included and all the adverse side-effects are carefully investigated. The overall criterion is to demonstrate that the new antibiotic actually has a role, preferably with clinical advantages over similar drugs currently available. The Food and Drug Administration usually take about a year to give the drug its licence and it is this approval that is the essential goal that all companies strive for. There are different licensing authorities in Europe and Japan, and these also have to be convinced before the antibiotic can be launched in these areas; however, the North American market is the one that all antibiotics must break into if they are to make a significant financial return. At the point of registration with the Food and Drug Administration the success rate is near 100%; the costs have been so great to reach this stage that only those antibiotics that are virtually guaranteed success are entered. It is not really like an examination, as there is continuous dialogue between the companies and the licensing authorities as to what will be required.

The reasons for the escalating costs are in Phase III, as this includes a large number of clinical investigators who will all require financial reward for entering patients on the trial. Actually they have to provide a large amount of data for each patient, which will include information on the efficiency of curing the symptoms, the eradication of the bacteria from the site of infection, the ease of administration and adverse effects. Each patient entered into the trial will cost thousands of US dollars just at the site of drug administration, let alone the administrative costs of the pharmaceutical company. Some years ago, a colleague in a major pharmaceutical company told me that his department reckoned that half of the clinical investigators recruited to enter patients on a Phase III trial either produced no data at all or gave insufficient information to be of value.

The major difficulty comes with the opportunity of the pharmaceutical company to recover its costs. Antibiotics, like all pharmaceuticals, are given a fixed patent life; the pharmaceutical company is given the sole marketing rights for only 25 years. The clock does not start ticking when the drug is launched but when the patent application is submitted. This is a very difficult equation for the company to balance; if it patents the drug as soon as it is discovered, the sheer volume of drugs at this stage will ensure that a large number of patents are filed and this would be very expensive. However, by the time the drug is launched, there may only be five years of patent life left. This is really insufficient to recover the costs. Most companies delay the filing of a patent until they are sure that the drug is likely to pass the early phases, though the overriding consideration is probably whether their competitors are close enough to discover the chemical structure of the new compound and so might steal the march.

The end of the patent life does not prevent the company from making substantial financial gains from their drugs; it is just that they do not have the sole rights and consequently they no longer determine the price. There are a number of companies, called me-toos, that specialise in the distribution of drugs that are out of patent. They do not even have to manufacture the drug, they just market it at a cost far lower than the original discovering company.

They have had none of the development costs so essentially just have to recoup the costs of manufacture and marketing. They are never allowed to use the marketing name, which remains the property of the discovering company; instead they use the generic name, which may deliberately be made awkward so that the original marketing name is the one that is generally remembered. This may have been the case for the combination of trimethoprim and sulphamethoxazole which had an original marketing name of Septrin, but its generic name is co-trimoxazole, which is far more cumbersome. If a medical practitioner prescribes an out-of-patent antibiotic and supplies the generic name on the prescription, the pharmacist may supply the generic drug from any manufacturer or may supply the original drug. On the other hand, if the prescription states the marketing name, this is what must be supplied. During the period that the company has the sole marketing rights, it will usually run a vigorous advertising campaign to ensure that the marketing name is one that is firmly placed in the mind of the prescriber rather than the generic, so when the drug comes out of patent, the familiar prescribing practices will continue. Actually this is a fairly good assumption; when the generic companies start competing, the original manufacturer often reduces the price to compete, maintaining a small differential. This is usually enough to keep product loyalty.

The determination of the patent life is a crucial part of the debate about the development of resistance because it dictates the marketing strategy of a new antibiotic. If we assume that a company patents a new antibiotic within two years of its discovery, by the time it reaches the marketplace there may be seven years left of patent life. A company would anticipate making a return on its investment before the patent expired. This would demand that the drug is introduced for the treatment of all likely infections as soon as possible rather than by the gradual introduction that might help to preserve its efficacy for longer. It also makes antibiotics that are designed for community infections much more attractive than those destined for nosocomial infections, which is where a dearth in new drugs is really beginning to be felt. With drugs other than antibiotics, deluging the market with a new product is not usually detrimental as resistance is never an important consideration, so the patent laws might work well with non-antibiotic pharmaceuticals.

The argument for maintaining the present patent laws are promoted by organisations such as the World Health Organisation who suggest that while a pharmaceutical is under patent, it is effectively excluded from the developing world and it might also be excluded from the poorer members of the industrialised countries. This does ignore the fact that some developing countries pirate new drugs almost immediately a new drug is launched in the west, but certainly the patent does deny access to pharmaceuticals. Bearing in mind that resistance development is often significantly faster in these medical environments, any practice that reduces the use of new antibiotics must help to preserve their efficacy for the whole global village. The use of pencillins was highly successful for the treatment of gonorrhoea in the west, but once these drugs were introduced for the treatment of sexually transmitted diseases in South-East Asia, there was an explosion of penicillin resistance which soon spread to the west.

There are some who argue that antibiotics should be treated differently from other pharmaceuticals and that the patent life should be lengthened. These arguments do not just emanate from pharmaceutical companies who would, of course, make more money from this. If the patent life was longer, most companies could attenuate their marketing campaign, releasing the drug in a manner appropriate for proper health-care management rather than rapid profit. It is often stated that gamekeepers make the best conservationists; although their jobs might appear to be to kill animals for sport, they actually have to preserve sufficient stocks of game animals. Similarly, the destruction of the efficacy of a new antibiotic, through the development of resistance, by a rushed marketing campaign is not in a pharmaceutical company's interests, but they have to pay a dividend to their shareholders. Most companies would prefer to recoup their costs for antibiotics over longer periods of time, as the longer the drug remains active, the longer they will make money from it. The reality is that the preservation of antibiotic efficacy and the reduction in the speed of resistance development is in everyone's interest. It is certainly good for the consumer, particularly faced with dwindling options, but it is also good for the discoverers of the drugs who are able to maximise their profits while minimising the detrimental impact on the patients. I suppose the main losers would be the me-too companies, but they have contributed the least so perhaps we should not be too concerned about them. I am not suggesting that the patent on antibiotics should be indefinite, rather that perhaps the clock should either start running once the drug is launched, or if this is considered to be unacceptable as it would mean that companies could hang on to new drugs until their previous compounds were coming out of patent, then the present system should continue but the patent life might be extended to 30 or 35 years.

These multinational companies are often seen a ruthless exponents of the capitalist ideal, with gross national products to which most countries of the world would dearly love to aspire. They are often seen as mercilessly dumping drugs that they cannot pass through western licensing authorities on the unsuspecting developing world. There may be some truth in this perception, with some companies; however, the competition that they have set up amongst themselves has proved to be the only method to discover and develop new pharmaceuticals. This is graphically seen when we examine how many new antibiotics were discovered and developed by the state-run pharmaceutical companies of the countries belonging to the former Warsaw pact; in truth they simply made clones of drugs that the western companies had pioneered.

PROSPECTS FOR NEW DRUG DEVELOPMENT

The solution to the rising problem of antibiotic resistance is either new antibiotics or completely different methods to deal with the problem of bacterial infection. We have just investigated the disincentives for discovering new antibiotics, especially as what is really sought is drugs that inhibit completely new targets; however, what are the realistic possibilities, bearing in mind that no new targets have been found for well over 30 years? Except for the discovery and development of trimethoprim, all antibiotics have been discovered by serendipity – the drug was discovered and then the target was identified afterwards; however, this was not a prerequisite to using the drug. This is a very simplistic approach to discovering new drugs but it still persists in most pharmaceutical companies; most retain a natural products division where soil samples are collected from around the world and tested for signal properties, including antibacterial activity. I have collected soil samples in the central Australian desert and on the Terai plain in Nepal with this aim in mind; the microorganisms never amounted to anything. The argument for maintaining this approach is that the natural products that became the first antibiotics formed the basis of the best antibacterial drugs that we have ever devised. A few clinical compounds are completely synthetic but, by and large, these have proved less sustaining than their counterparts derived from natural products. In reality, as we face a treatment crisis at the start of the new millennium, new treatments from any source would be welcome.

A molecular approach to discover new antibiotics is being initiated by many of the largest pharmaceutical companies. The human genome project has gained a considerable degree of publicity, and laboratories all over the world are investigating the sequence of the total DNA in all 46 of the human chromosomes. Similarly, there are laboratories which are determining the DNA sequence of the single chromosomes of pathogenic bacterial species. The first to be completed was the sequence of the respiratory pathogen, *Haemophilus influenzae*. This organism is fairly simple in bacterial terms and the chromosome was rather smaller at 1.8 million nucleotide bases than many other species such as *Escherichia coli*, which might average above 4.6 million. The sequence of *Haemophilus influenzae* is published and in the public domain; information on it can be found on the Internet at TIGR at http://www.tigr.org. Like the human genome project, the bacterial chromosome project is plagued with self-interest and some sequences have been completed and patented by some companies. I do not know much about patent law but I have difficulty in understanding how a sequence can be patented; it would seem that the structure of a living organism is in the public domain and the information on it should be freely available. Nevertheless, there are some obvious exceptions in the list of pathogenic bacteria in the TIGR list and some of these have been done but are patented.

The determination of a complete bacterial DNA sequence might seem an unlikely method to find a new antibiotic. If the structure of the complete genome is known, then target genes can be identified. The identification of suitable target genes will come from searches of the literature, which might suggest which are required for the integrity of the cell. It is possible to disable the gene selectively and find out whether the lack of this gene's function is lethal to the cell, and it is unlikely that any drug would now undergo serious development unless it was shown to kill the bacterium. The structure of the gene will be compared with genes from mammals and particularly humans. If there are equivalent human genes, comparative analysis would be performed to see if they are closely related. This might show if an inhibitor of the bacterial gene product would be likely to have a similar inhibitory effect in human cells. If there is some diversity between the bacterial gene and the human equivalent, then selective inhibition of the bacterial gene production might provide sufficient selectivity.

The other consideration is that if we are able to produce designer drugs, they should be active against certain pathogens. We have seen how many of the major problems in hospitals are caused by multi-resistant gram-positive bacteria; it should be possible with this approach to identify suitable genes that are prominently in gram-positive bacteria and are less evident in gram-negative, thus leading to the development of antibiotics that would be active specifically against these bacteria. Once a prospective gene has been identified, comparison of the genes of related bacteria would suggest whether an inhibitor of this gene would have a broad spectrum or whether it would be confined to a narrow range of pathogens.

Once a suitable gene has been identified, it can be selectively cloned in a known bacterium so that its function can be analysed in detail under defined conditions. A single gene usually encodes a single protein which can be studied in greater detail if the gene is cloned and cloning will be relatively easy if the whole genome sequence is known. The gene will be excised with suitable restriction endonucleases; these enzymes have recognition sites composed of very specific sequences so examination of the flanking DNA sequences of the gene will reveal which restriction endonuclease will optimally excise the gene; this is usually a rate limiting step in the cloning of genes. Cloning of the gene into a suitable vector plasmid will permit an escalation in the production of the gene's protein product. The protein can be purified and its 3-D structure determined by X-ray crystallography. The amino acid sequence of the protein can be deduced from the DNA sequence and thus the exact spatial structure of the protein and its binding sites can be established, and this information, as a 3-D model, can be modelled in a computer. The computer can then be used to devise molecules that will bind to this virtual protein. The effects of altering chemical functions can be found in minutes where before it would have taken months for chemists to devise the compound and then it would have been tested. When a suitable virtual inhibitor has been identified, chemists can be given the task of synthesising it. After synthesis, microbiologists and biochemists can verify whether this compound really does inhibit the protein. This approach is very much cheaper and more efficient than the one currently adopted of the continual synthesis of new chemical compounds which are painstakingly tested, the vast majority of which never even enter Phase I.

I believe that there is a central flaw in this approach as it assumes that all strains within a species are identical or at the very least sufficiently similar that the differences between them are insignificantly small that it would make no difference.

The pharmaceutical companies have undertaken a massive campaign to sequence complete genomes and spent millions of dollars doing so. The sequences are patented so that the pharmaceutical companies can use this information exclusively in the design of new drugs. As sophisticated as these sequencing techniques are, they are in stark contrast with the methods that we use to identify bacterial species. Bacterial identification is still largely based on methods that were initiated a century ago and rely on the biotype of the strain; that is to say they are dependent on the bacterium to produce certain enzymes whose function can be identified by simple biochemical reactions. This system identifies *Escherichia coli* as a lactose-fermenting gram-negative rod whereas *Salmonella* species, which are virtually identical, do not ferment lactose and appear quite different in these tests. There are certainly some variants of *Escherichia coli* that do not ferment lactose, we should really not call these *Escherichia coli*. It raises the issue of how similar are strains that are identified within the same species. *Escherichia coli* is a good example; for all intents and purposes there is only one species throughout the world but there must be enormous variation between *Escherichia coli* identified in one region compared with another. We have examples of this variation in our own laboratory. We have looked at the *gyrA* gene, the primary target of ciprofloxacin, of this species identified by conventional tests isolated in Europe and the Thailand–Malaysian peninsula. There are certainly differences between the genes that are not just related differences in response to antibiotic challenge. These changes are almost certainly due to indigenous differences between the genes and reveal that the genes and the host bacteria are fundamentally different. Therefore, if we were to sequence the genome of *Escherichia coli* from Europe, perhaps from Edinburgh, and then, based on this information, we were to design a new drug that should be a good candidate to inhibit the *Escherichia coli* *gyrA* gene, when this drug was used to treat infections in South-East Asia, it *might* not be effective.

This type of diversity becomes apparent when considering the emergence of an enzyme like the BIL-1 transferable β-lactamase, which we identified in *Escherichia coli* in 1990. The gene was sequenced and found to be most similar to the chromosomal β-lactamase gene of *Citrobacter freundii*. The degree of this similarity was about 93% and for the most part, the gene shared similar biochemical properties with the *Citrobacter freundii* enzyme except in the manner in which it hydrolysed cephalothin. In this respect it was more similar to the *Enterobacter cloacae* and *Escherichia coli* chromosomal β-lactamases. However, the sequence similarity suggested that it was a *Citrobacter freundii* enzyme that had migrated from the chromosome and while moving through the clinical population on its plasmid harbinger, the gene underwent a series of mutations that befitted it for the needs of the host strains. This is quite a conventional hypothesis for the development of resistance genes but it does assume that there has to be a 6% variation from the original *Citrobacter freundii* gene. It is well documented that such variation can occur in a limited number of genetic movements, but this is by the system of "mosaic" transitions, normally found in a few gram-positive species, which involves the movement of complete blocks of the gene and substitution of this new block in the original gene. Thus when the new gene is analysed, there will be blocks of the old gene interspersed with blocks of new gene "mosaic". The BIL-1 gene does not follow this model; the variation from the *Citrobacter freundii* gene is much more random and does not come in blocks. Whereas the BIL-1 β-lactamase gene and the *Citrobacter freundii* must come from a common source, it does not necessarily follow that one was derived from the other. Indeed, it is probably unlikely that this happened. It is convenient to suggest that the BIL-1 gene underwent many mutations; however, this is very unlikely. Although mutations can occur quickly in bacteria, they are often detrimental and many random mutations probably have to take place before a successful mutation emerges because it confers sufficient advantage on the host. This is a lengthy business and I think impossible in the time frame since antibiotics were first used clinically. We only have to examine the extended-spectrum β-lactamase; they are best example of resistance mutations emerging in response to antibiotic challenge. The maximum number of mutations that occurred within the antibiotic era was four, much less than 1% of the whole gene. There simply is not enough time for the required number of mutations to occur, and then be successful.

A discussion between Nancy Hanson of Creighton University and I, led to a much more likely explanation is that the gene actually came from a *Citrobacter freundii*-like strain. It might even have been classified as a *Citrobacter freundii* if it had been examined by conventional biochemical tests. The original host was probably related to *Citrobacter freundii* and may or may not be a bacterium found in the clinical environment. The problem is that if the strain would be identified as a *Citrobacter freundii* by conventional testing, then any drug that had been developed to inhibit the β-lactamase of the prototype *Citrobacter freundii* may well not inhibit the enzyme from the closely related species.

Bacterial speciation is a system that we have developed to make it easier for us to identify the cause of infection. Actually, with the advent of antibiotics, accurate identification might not be essential if the infection can be controlled, but if we are to base selective toxicity data on this information then it is essential that it is not only accurate but also sensitive enough for modern needs. The systems devised for bacterial identification are not sufficiently sensitive and it is imperative that we devise a system based on genetical or molecular biological criteria. As computers become more sophisticated, it will be possible to analyse the whole genome of a microorganism in a reasonably short time. We are not at that stage at the moment and we can only approximate our genetical analysis of bacteria by some arbitrary molecular technique. Various procedures have been used including RAPD (Random Amplified Polymorphic DNA) which often uses a single oligonucleotide in a PCR tube which will bind to the genomic DNA and start to amplify the sequences in between the points of binding. A number of different-sized fragments will be generated and these can be separated in a conventional agarose gel. As the points of binding should be constant, this type of random amplification should produce a characteristic pattern and, within the same laboratory, it often does and this technique can be used to compare bacteria of the same species. It does not, however, give any indication of the degree of similarity; it merely shows whether one strain is closely related to another. The problem with this technique is that amplification in a PCR tube is very dependent on the conditions used and so the resultant pattern varies with each different machine and patterns are not consistent. It is not usually possible to compare information obtained in one laboratory with that obtained in another. A more useful technique is to analyse the genomic DNA directly. Very accurate fingerprinting of plasmid DNA can be obtained by isolating the DNA and cutting it with restriction endonucleases. As the recognition sites of restriction endonucleases are very specific, the exposure of DNA to these enzymes will form fragments of fixed size. These can be separated by agarose gel electrophoresis and will give a characteristic pattern. This pattern is unlikely to be variable as the restriction endonuclease reaction is less dependent on alterations of conditions, unlike the polymerase chain reaction. This system works extremely well with clinical plasmids with sizes up to 150 kb and, in the absence of plasmid sequencing, is the most accurate method for comparing plasmids. The technique can be applied to whole bacterial genomes but unless enzymes with relatively rare recognition sites are used too many fragments are generated. Thus enzymes that require six or more bases for recognition are imperative. The problem with the use of such rare "cutters" is that the fragments they generate are much larger; they may be more than 1000 kb. If these fragments were separated in conventional agarose electrophoresis, the large fragments would bunch up near the well as they would hardly migrate into the gel.

This problem has been overcome by altering the conditions of separation of the fragments. Instead of applying a conventional electrical field, the electrical field is varied; it is "pulsed" so that the fragments are sent to one corner of the gel, and the field is then changed so that it is then sent to the opposite corner of the gel. The conditions of the pulsing can be changed according to the conditions needed but the net effect is to tease the large fragments away from one another. The overall migration is straight down the gel, which is similar to conventional electrophoresis but the large fragments are separated one from another. This produces a very characteristic pattern which is consistent and would be similar for the same organism if it was performed in my laboratory or in a laboratory on the other side of the world. So consistent is this technique that the image can be scanned into a computer and stored in a database. Indeed there are now databases of some bacteria to which this type of information can be entered.

This information does mainly show differences and not similarity. There have been algorithms written that estimate, based on the likelihood of changes to the restriction sites, the similarity of two species to one another. Indeed, quite a number of computer-illiterate microbiologists use this information as factual to make quantifiable deductions of similarity. We have already seen how misinterpretation of resistance gene information has greatly skewed our view of what is happening in the spread of some resistance genes. So too, the unquestioning use of computer algorithms may also bias our view of the spread of certain microorganisms.

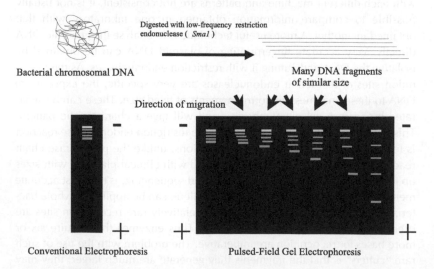

Pulsed-Field Gel Electrophoresis

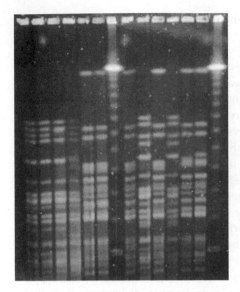

No	Strain No	PFGE	Source
1	1609	1	CEMIC - catheter
2	1610	1	CEMIC - catheter
3	1611	1	CEMIC - catheter
4	1613	1	CEMIC - catheter
5	1614	2	CEMIC - BAL
6	1615	2	CEMIC - BAL
7	1616	2	Teaching hospital - BAL
8	1617	3	Teaching hospital - BAL
9	1618	2	Teaching hospital - Soft tissue
10	1619	3	Geriatric hospital - Urine
11	1622	2	CEMIC - BAL
12	1623	2	CEMIC - BAL

Pulsed-Field Gel Electrophoresis of ARI-2 β-lactamase-producing strains from Argentina

What can we interpret from a Pulsed-Field Gel Electrophoresis pattern? If two bacteria have their genomic DNA extracted and then cut with the same restriction endonuclease, so that they produce identical patterns when separated by electrophoresis, we can interpret from this that they are closely related. They may be identical but we cannot interpret this from the data because the information presented to us relies on the conservation of certain restriction endonuclease restriction sites. If some of the banding patterns are different, the bacteria are often assumed to be similar, though this cannot be quantified and *may* or *may not* be significant. It merely demonstrates that some of the restriction sites are different. Some microbiologists state that if three or more banding patterns are different then the bacteria are different, though it is often unclear what they mean by different. If there is one band change the bacteria are different, as they cannot be identical. Again it is impossible to quantify this difference. I think that attempts to demonstrate identity might be misplaced though some interpretation of similarity could be inferred. This should probably be a qualitative interpretation rather than a quantitative interpretation.

The reader may wonder what this has to do with the identification of new antimicrobials. We merely need to know, if we are devising a drug to act against a specific microorganism, whether the specificity of our interpretation of the identity of that bacterium is sufficient to guarantee that the new drug will attack all members of that species, or whether the variation within that species is so great that only some members will be inhibited or only those within a specific locality.

Perhaps the most concerning variation amongst a single species is in *Staphylococcus aureus*. This species has always caused significant clinical problems from the moment that it was first identified. It became more problematic when it first became resistant to the penicillins in the mid-1940s, to methicillin in the early 1960s and to both methicillin and gentamicin in the 1980s. As methicillin resistance developed, it was assumed that previously sensitive bacteria had acquired methicillin resistance. The methods used to identify and type *Staphylococcus aureus* were archaic; it was still often identified by the production of golden colonies and typed by its relative sensitivity to a range of bacteriophages. Unfortunately, when the bacterium acquired resistance to methicillin, it was not easily lysed by bacteriophages and it was assumed that the appropriation of methicillin resistance, in some way, destroyed the ability to bind the bacterial viruses. The belief persisted that what was being observed was resistance emerging in the existing hospital staphylococcal population; however, as typing procedures improved, particularly with the advent of molecular techniques, it became clear that epidemic strains of MRSA were spreading. These bacteria were not only able to move between hospitals within a city but even to spread between continents. The genomes of some of these staphylococci are currently being sequenced; it is not known how closely related the early *Staphylococcus aureus*, such as the Oxford staphylococcus, are to the epidemic MRSA strains EMRSA-15 or EMRSA-16. It has been suggested by molecular taxonomists that if they share 70% sequence homology, then they are likely to be the same species. This is a totally arbitrary distinction which *may* or *may not* give any indication of similarity. If that criterion was used amongst the Enterobacteriaceae, then some *Escherichia coli* and many *Salmonella* species should be considered as the same species. It does have two important implications; firstly, if decisions are to be taken to develop a new antibacterial which are based on comparative genome sequence analysis, it is essential that the correct strain is sequenced. I suspect, for instance, that relatively little could be deduced from the sequence of the Oxford staphylococcus that would be relevant to the selective destruction of the epidemic MRSA in the United States, for example. The other difficulty comes in the patentability of a sequence. We are considering variations within a species group that are far greater than the relatively minor variations that occur within *Homo sapiens*, where patenting of genome sequences is also a contentious issue. When we consider a single species, which appears to be the criterion for the consideration of a patent, then if the sequence of the Oxford staphylococcus is sequenced and then patented, does this mean that the sequencing of the EMRSA-16 is not permitted or, if it is, can that new sequence be published and entered into the public domain and thus be available for other scientists to use? These questions have not yet been answered, nor tested properly in the courts. It is my contention that no sequence information should be patentable, bacterial or human.

These discussions do demonstrate, however, that our definition of species have become antiquated and we need far better definitions of species, probably based on molecular criteria. These probably cannot be based on techniques such as pulsed-field gel electrophoresis because these results are not quantifiable, but rather they should be based on sequence data as this becomes available. Perhaps before we start looking for new antibacterials, we should seek the extent of the problem and really try to identify the exact nature of the pathogens that we are facing. We should determine exactly what type of pathogens we are confronting; are the staphylococci in London the same as those infecting patients in Durban? If they are not, are the strategies that we should use to combat them the same? The idea of molecular modelling and how drugs might bind to specific targets is exciting; however, it relies entirely on the integrity of the information provided. If this information is flawed in any way, then data generated based on this information will exacerbate this flaw. If the wrong pathogens are sequenced and new antibacterials are based on this information, the chances of ultimately finding a successful antibiotic are diminished.

Spread of bacteria – resistant or sensitive!

ANTIBIOTICS FOR A NEW MILLENNIUM

As we enter a new millennium, we do not have a batch of new antibiotics to cope with the infections that we have to face in the future, particularly those that have either acquired resistance or those resistant species that we do not yet know about. If we do not have new antibiotics, what are our likely prospects? What is presently declared by the pharmaceutical companies about their prospective antibiotics? There are some new variants of the fluoroquinolones, particularly with increased activity against gram-positive bacteria. They have been developed for their activity against *Streptococcus pneumoniae*. These drugs might have slightly enhanced activity against some MRSA, but would we really believe that this most adaptable group of bacteria is incapable of dealing with new variants of the fluoroquinolones? It would very surprising if they were not. The enterococci are virtually unaffected by these new fluoroquinolones. We enter the new millennium in the developed world with three hospital bacteria, some strains of which are virtually untreatable with conventional antibiotics: vancomycin-insusceptible, methicillin-resistant *Staphylococcus aureus*, vancomycin-resistant enterococci and carbapenem-resistant *Acinetobacter baumannii*. What will that mean to hospitals in the future?

Methicillin-resistant *Staphylococcus aureus* and *Acinetobacter baumannii* have been responsible for closure of hospital wards, on both sides of the Atlantic, while they have been fumigated. These bacteria were not even the most resistant variants of these species and these closures caused significant decreases in the provision of health care. The totally resistant bacteria are unlikely to go away spontaneously. There used be an old-wives tale associated with antibiotic resistance: if you did not use an antibiotic for a while, resistance to that drug would soon disappear. This is clearly not the case and we can expect some of these totally resistant bacteria to persist for some considerable time, even if we no longer use the antibiotics to select them. Another resolve of microbiological folk-lore is that resistant bacteria are far less virulent than their sensitive counterparts. This belief came from the selection of resistant bacteria in laboratory experiments. The bacteria selected in overnight mutation experiments had often had to make such radical changes to survive in the new antibiotic environment, perhaps by making an alteration that was expensive in energy terms, that the production of proteins involved in the pathogenicity of the bacterium was compromised. This is a situation that is unlikely to occur in the clinical situation; after all, we isolate these multi-resistant bacteria precisely because they are pathogenic. These are identified in the usual sites of infection so the general premise of this belief is probably unsound.

What will happen as we enter the new millennium with these totally resistant bacteria, which are probably as pathogenic as they have always been? We have already seen that the acquisition of vancomycin resistance for enterococci has helped raise their profile so that these bacteria are now the second most common pathogen in Intensive Care Units. Dr. Susan Brown in my laboratory has recently identified an epidemic carbapenem-resistant *Acineotbacter baumannii* harbouring the ARI-2 β-lactamase; this totally resistant strain is already responsible for infections on four continents. This has helped raise *Acinetobacter baumannii* to become the most problematic gram-negative pathogen in the Intensive Care Unit. These are likely to be the two major problems in the early years of the new millennium. Recently, I took part in a video that suggested that totally resistant enterococci will radically affect our confidence to perform medical procedures in the near future. This might be considered alarmist but we must consider the implications. If we have untreatable bacteria capable of infecting patients in Intensive Care Units, then a new equation will have to be solved before the decision to initiate certain procedures is taken. At the moment, when a patient undergoes transplant surgery various considerations have to be taken into account; what will be the potential benefit to the patient and will this outweigh the risks? If the chances of successful surgery are high and the risks of damaging bacterial infection are low, provided a suitable donor organ can be found, the decision to proceed is quite straightforward even if the patient would have just a few years of added benefit. Depending on the procedure, the patient might not live longer but the quality of life might be considerably increased.

How would this equation balance if the risks of damaging infections were high? For this type of surgery, the patient has to be immunosuppressed and therefore, must have some antibiotic cover to prevent the inevitable bacterial infection. Suppose the chances of uncontrollable infection were 45%; this may lead to septicaemia and might result in an overall mortality rate of perhaps 30%. Would a 30% mortality rate be acceptable for an improvement in the quality of life? That is a question that I cannot answer but may well have to be posed to individual patients contemplating certain types of elective surgery. We have no drugs in the foreseeable future that might be able to alleviate this situation.

The number of patients that end up in Intensive Care Units is a relatively small proportion but is currently increasing. Enterococci and *Acinetobacter baumannii* are currently not problematic in general hospital wards. The latter is a common cause of respiratory infection in hospital but might not be a problem in patients who do not require ventilation. *Staphylococcus aureus* does most damage outside the Intensive Care Units and, for this reason, is likely to be the most damaging hospital pathogen that we face. If we have totally resistant strains it is going to compromise many medical procedures and these will not be confined to the seriously ill. It may not be difficult to weigh the consequences of infection against the benefits of surgery when the condition is life-threatening. When the procedure is to relieve suffering or even cosmetic, the risks imposed by infection are much greater. We could take the example of hip-replacement. This is a procedure that is often offered to arthritis patients, particularly those in the latter half of their lives. It never saves a life, rather improves the quality of it. Because of the usual age of the patients, the effects of a hip-replacement may not provide increased mobility for very long periods of time. When a surgeon consults a patient for a hip-replacement, he or she may outline that there is a risk of infection as they describe the procedure. The patient might also be told that this risk might be as great as 50% but it is unlikely that the patient will be told that there is a significant risk that this infection will carry the risk of death. The reason is because the patient is protected with antibiotics. If the hospital is infiltrated with vancomycin-resistant *Staphylococcus aureus*, then the counselling of the patient will have to take a very different character. Although the risk of infection may be no greater, the risk of serious consequences is very much greater because if the bacterium were to cause septicaemia, it would be untreatable. It is impossible now to assess what that risk is. We currently cover our patients and it might be considered unethical if we did not. If the risks of this surgery were that 40% of the patients were infected with a totally resistant epidemic *Staphylococcus aureus*, and the potential mortality from this is 25% (it could be higher), this would be considered a totally unacceptable risk for a condition that was not life-threatening. What would the patient be told? The patient might be informed that the surgical procedure was available and that it would, if successful, improve the patient's quality of life. In a hospital that did not have a totally resistant epidemic *Staphylococcus aureus* then the chances of success might be greater than 90% and failure would not be likely to be life-threatening. There are many patients who would accept this risk for an improved quality of life. On the other hand, in a hospital riddled with a totally resistant epidemic *Staphylococcus aureus*, the patient should be told that the odds are rather different.

The problem will be severe with the three totally resistant pathogens described above, especially with *Staphylococcus aureus*. It will be further exacerbated if other pathogens become totally resistant. *Klebsiella* species are gaining increasing importance as they become multi-resistant. These pathogens are ubiquitous and can infect many sites. In the last few months of 1997, there were increasing reports of imipenem-resistant *Klebsiella* strains isolated in hospitals around the world, including the United Kingdom. Multi-resistant, but treatable, variants of this genus caused much concern in hospitals in the 1970s and early 1980s, although the problem can be contained. It is possible that *Pseudomonas aeruginosa* will also become totally resistant, as there are many reports of carbapenem resistance. It appears to be a less virulent pathogen in hospitals than it has been in the past. *Burkholderia cepacia*, a bacterium that infects patients with cystic fibrosis, is now almost totally resistant, leaving infected patients unprotected. It can also attack other patients in hospital. The causative agent of pneumococcal pneumonia, *Streptococcus pneumoniae*, is becoming significantly resistant to penicillin in some parts of the world; if this species acquires as many resistance genes as *Staphylococcus aureus*, we are likely to face a severe crisis in the community, particularly with the elderly. *Mycobacterium tuberculosis* is also in danger of becoming resistant to all antibiotics that are used to treat it in certain parts of the world. This would be a total disaster for whole communities, particularly in poorer regions of the world. It could destroy whole countries and this would be a much greater crisis than AIDS.

CONCLUSION

This story is still continuing so it might be inappropriate to call this section a conclusion. It would be more appropriate to call it a cross-roads. What happens in the next five years is going to be crucial. If either no further innovative antibiotics are released for severe hospital infection or we do not take measures to curb the spread of resistance, then we are going to slip further into an abyss of uncontrollable infection. It will, of course, be possible to reverse that trend should a new drug be found at a later date; but in the meantime, many people will succumb to and be killed by infection needlessly, because more judicious use of our antibiotic resources that we had available to us would have made them sustainable. It is possible that we never find another major class of drugs, though I hope this is unlikely; however, if this does occur we shall revert to a dark age of medicine. Which it will be, none of us can predict. All we can predict is that unless new antibiotics are found or we can find alternatives, the future could be very bleak.

Index

255